Current Topics in Microbiology
125 and Immunology

Protein Secretion and Export in Bacteria

Edited by
Henry C. Wu and Phang C. Tai

With 34 Figures

Springer-Verlag
Berlin Heidelberg NewYork Tokyo

Henry C. Wu, M.D. Ph.D
Professor of Microbiology
Uniformed Services University
of the Health Sciences
4301 Jones Bridge Road
Bethesda, MD 20814-4799, USA

Phang C. Tai, Ph.D.
Senior Staff Scientist
Department of Metabolic Regulation
Boston Biomedical Research Institute
20 Staniford Street
Boston, MA 02114, USA
and
Associate Professor
Department of Microbiology
and Molecular Genetics
Harvard Medical School
23 Shattuck Street
Boston, MA 02115, USA

ISBN-13:978-3-642-71253-1 e-ISBN-13:978-3-642-71251-7
DOI: 10.1007/978-3-642-71251-7

Softcover reprint of the hardcover 1st edition 1986

Library of Congress Catalog Card Number 15-12910

2123/3130-543210

Preface

The last decade has witnessed rapid progress in our understanding of the mechanisms of protein export and secretion in both prokaryotic and eukaryotic cells. Studies of protein secretion across the membranes of the rough endoplasmic reticulum have led to the formulation of the now-classic signal hypothesis, which has stimulated many discussions and new ideas, and the identification of the signal recognition particle as an organelle in the initiation of the export process. However, more recent work pertaining to intragenic information related to targeting specific proteins for either secretion or membrane localization, the energetics of protein secretion, the timing of synthesis versus the initiation of export, structural requirements for the processing of precursor proteins, and the identification of the processing enzymes (signal peptidases), has been the result of a combined biochemical and genetic approach to the study of protein localization in bacteria.

While reviews on the biochemistry and genetics of protein secretion have appeared frequently in recent years, this book attempts to summarize the current status and the future perspectives of this rapidly moving field in a single volume. Topics covered in this book include the genetics of protein secretion in *E. coli*, biochemical analysis of protein export in vitro, signal peptidases, excretion of colicins and hemolysin in *E. coli*, protein secretion in *Bacillus*, and protein secretion cloning vectors. Approaches encompassing classical bacterial genetics, membrane biochemistry, bioenergetics, and recombination DNA technology are utilized in a concerted effort for the elucidation of the molecular mechanisms of protein secretion in bacteria. It is our hope that this book will be valuable, not only to those actively engaged in studies of protein secretion in bacteria, but also to those who are interested in protein targeting and protein trafficking in general.

We would like to thank the staff of Springer Verlag for their help in the production of this book.

Henry C. Wu
Phang C. Tai

Table of Contents

Indexed in Current Contents

List of Contributors

BASSFORD Jr., P., Department of Microbiology and Immunology, University of North Carolina, School of Medicine, Chapel Hill, NC 27514, USA

BECKWITH, J., Department of Microbiology and Molecular Genetics, Harvard Medical School, Boston, MA 02115, USA

BLOBEL, G. Laboratory of Cell Biology, The Rockefeller University, New York, NY 10021, USA

DAVIS, B.D., Bacterial Physiology Unit, Harvard Medical School, Boston, MA 02115, USA

DEV, I., Department of Microbiology, Wellcome Research Laboratories, 3030 Cornwallis Road, Research Triangle Park, NC 27709, USA

FERRO-NOVICK, S., Department of Microbiology and Molecular Genetics, Harvard Medical School, Boston, MA 02115, USA

DE GRAAF, F.K., Department of Microbiology, Biological Laboratory, Vrije Universiteit, De Boelelaan 1087, NL-1081 HV Amsterdam

GRAY, L., Department of Genetics, University of Leicester, University Road, Leicester, LE1 7RH, United Kingdom

HOLLAND, I.B., Department of Genetics, University of Leicester, University Road, Leicester LE1 7RH, United Kingdom

INOUYE, M. Department of Biochemistry, State University of New York, Stony Brook, NY 11794, USA

LUNN, C.A., Department of Biochemistry, State University of New York, Stony Brook, NY 11794, USA

MACGREGOR, C., Department of Microbiology, Georgetown University Medical School, Washington, DC 20007, USA

MACKMAN, N., Department of Genetics, University of Leicester, University Road, Leicester LE1 7RH, United Kingdom

MÜLLER, M., Biochemisches Institut der Universität Freiburg, Hermann-Herder-Str. 7, D-7800 Freiburg

NICAUD, J.-M., Department of Genetics, University of Leicester, University Road, Leicester LE1 7RH, United Kingdom

OUDEGA, B., Department of Microbiology, Biological Laboratory, Vrije Universiteit, De Boelelaan 1087, NL-1081 HV Amsterdam

RAY, P., Department of Microbiology, Wellcome Research Laboratories, 3030 Cornwallis Road, Research Triangle Park, NC 27709, USA

SARVAS, M., National Public Health Institute, Mannerheimintie 166, SF-00280 Helsinki 28

TAI, P.C., Department of Metabolic Regulation, Boston Biomedical Research Institute, Boston, MA 02114, USA and Department of Microbiology and Molecular Genetics, Harvard Medical School, Boston, MA 02115, USA

TAKAHARA, M., Biochemical Research Laboratory, Tokyo Jozo Co. Ltd., 632-1 Mifuku, Ohito-Cho, Tagata-Gun, Shizuoka-Ken, Japan 410-23

TOKUNAGA, M., Mitsubishi-Kasei Institute of Life Sciences, Tokyo 194, Japan

WICKNER, B., Department of Biological Chemistry, University of California, Los Angeles, CA 90024, USA

WU, H.C., Department of Microbiology, Uniformed Services University of the Health Sciences, 4301 Jones Bridge Road, Bethesda, MD 20814-4799, USA

Past Triumphs, Future Challenges

B.D. DAVIS

The study of protein export, into or across membranes, has clearly become one of the liveliest areas in cell biology today, with rapid advances being made both in prokaryotes and in eukaryotic cells. But as a microbiologist, aware that the simplicity and the ease of genetic manipulability of prokaryotes made them the organisms of choice in the early explorations of many of the other universal properties of cells, I have been intrigued by the fact that studies of membrane structure and function have always lagged in prokaryotes compared with eukaryotes.

It may surprise students today that, for example, until the 1950s most microbiologists doubted the possibility, in a cell as small as a bacterium, of a selectively permeable membrane with a variety of specific transport systems. Thus in the first International Congress that I attended – Biochemistry, in Paris in 1952 – Krampitz presented excellent isotopic and enzymatic evidence for the presence of the Krebs tricarboxylic acid cycle in *Escherichia coli*; but Sir Hans Krebs himself, chairing the session, was skeptical, because the intact cells could not utilize exogenous citrate. And a decade later, Monod, finding that *E. coli* cells could concentrate a lactose analog, hesitated in his initial paper to suggest that this effect might be due to active transport (like that already known in animal and plant cells) rather than to specific adsorbing units within the cell.

Studies of the aspects of membranes taken up in this volume similarly lagged in prokaryotes. One reason was difficulty in visualization. Ribosomes are tightly packed in bacteria, and so a decade after Palade had demonstrated membrane-bound ribosomes in liver cells and postulated that they function in protein secretion, it was still not clear whether any of the ribosomes in bacteria were bound to membranes. Another reason was that the cytoplasmic membrane in the simple prokaryotes is, paradoxically, particularly complex, for it combines many functions that are divided among differentiated membranes in eukaryotes.

In support of the Palade model, Redman and Sabatini soon presented evidence that incomplete proteins, released by puromycin, could be secreted into the lumen of the endoplasmic reticulum. However, the study of protein secretion then lagged for nearly a decade. What rearoused interest was the discovery that the membrane-bound ribosomes do not differ from other ribosomes (as was at first assumed), but are attached to the membrane by a hydrophobic signal sequence on their nascent polypeptide. This possibility was suggested by Sabatini and Blobel in 1971; but if they had tried to test their hypothesis,

Bacterial Physiology Unit, Harvard Medical School, Boston, MA 02115, USA

Current Topics in Microbiology and Immunology, Vol. 125

using the then still arduous methods of protein sequencing to seek a hydrophobic initial sequence on the secreted protein, no such sequence would have been found. In a classical example of serendipity, the key was revealed in 1972 in an immunological study in which Milstein and his colleagues first achieved synthesis of an immunoglobulin chain by cell extracts (from a myeloma cell line). They unexpectedly found that synthesis in the absence of membrane (but not in its presence) yielded a product with an extra segment of ca. 3000 daltons; because this segment was N-terminal they proposed that it served as a signal, initiating entry into the membrane and then being cleaved by a membrane enzyme.

Schaechter showed that this signal segment was highly hydrophobic, and then Blobel and Dobberstein determined its sequence. The field was broadened when Blobel further showed that integral membrane proteins, as well as secreted proteins, are synthesized as precursors carrying such a cleavable sequence; he also achieved protein translocation into vesicles in extracts. Since no systematic differences have been found between the signal sequences of membrane proteins and those of secreted proteins, one of the major challenges in the field is to identify what later sequences determine this choice. A closely related problem is to determine what tickets various proteins for entry into various membranes. Here the signal sequence appears to play a role.

Bacteria, brought into the field of protein secretion belatedly, have led to two distinctive contributions. Cotranslational secretion was directly demonstrated by extracellular labeling or cleavage of growing chains – a conclusion also established in animal cells by demonstrating attachment of oligosaccharides, in the lumen of the endoplasmic reticulum, to growing chains. And extensive genetic studies, mostly by the Beckwith school, have illuminated several aspects of protein translocation. Mutations in the hydrophobic signal sequence have not only confirmed the expectation that insertion of polar residues would impair secretion, but also revealed more subtle relations between structure and function. More interestingly, unexpected fates have been observed for β-galactosidase fused with various lengths of proteins destined for secretion or for translocation to the outer membrane, and other genetic studies have also identified several *sec* genes whose mutations interfere with secretion. We might further note that the outer and the inner membrane in gram-negative bacteria provide a relatively simple example of a multiple-membrane system, whose mechanism(s) of distribution of proteins may well prove to be similar to those in their evolutionary descendants, the mitochondria and the chloroplasts.

In an important further insight into the steps that initiate export, Walter and Blobel have shown that some eukaryotic systems involve a nucleoprotein signal recognition particle on the ribosome, which binds the signal and arrests further synthesis until the ribosome interacts with the membrane. Some bacterial systems involve a simpler secretory particle on the ribosome, whose action is less clear. These findings are no doubt pertinent to cotranslational export. But after the triumph of establishing this process, the big surprise was the finding that it is not universal: some proteins are evidently secreted post-translationally, since they are made in the cell on unbound polysomes. Moreover, even proteins that are synthesized in the cell only on membrane-bound ribosomes can be inserted post-translationally into vesicles in extracts; and it turns out that many proteins are exported without the help of a cleavable signal sequence.

We are thus faced with multiple mechanisms. The most challenging problem is the mechanism that takes up a protein on one face of a membrane and releases it on the other. Though this process utilizes energy, how it does so is completely obscure. Moreover, while a pore through a contractile protein machinery initially seemed an attractive way of accounting for this feature, it seems increasingly likely that the protein traverses the membrane in a way that allows highly hydrophobic segments, when present, to interact with and be anchored in the lipid surroundings.

But now I am poaching on the papers in this volume, which cover the current state of the field. Let me close by noting that passage across a double membrane does not necessarily involve secretion across either component. With gram-negative bacteria there is good evidence that some excreted proteins do not pass successively through inner membrane, periplasm, and outer membrane but instead flow along the junctions from the inner to the outer membrane and then are released. A similar lateral flow may well occur in the passage of cytoplasmic proteins into the matrix of the mitochondrion. To avoid a prejudicial terminology, it might be useful to restrict the use of the term secretion to those instances where it is clear that the protein crosses a membrane and then is released to the aqueous phase on the other side.

I would also like to add a speculative suggestion: that the predominance of aqueous environments in living cells, and in the world that we inhabit, has probably exaggerated our impression of nonspecificity in hydrophobic reactions. In the interior of the membrane, the lipids, and hydrophobic surfaces of proteins, may well interact with much specificity, while intruding polar regions of proteins and polysaccharides would aggregate nonspecifically to escape the hydrophobic environment. This area of biochemistry is now at a stage akin to that of protein chemistry when I was a student in the laboratory of E.J. Cohn in the late 1930s. It was then believed that the behavior of different proteins could be best explained in terms of their electrical properties. Specific sequences, three-dimensional structures, and surface patterns were then out of reach and, indeed, beyond imagination; but they are accessible today. Similarly, specificity in the interactions of lipids with each other, and with proteins, is beginning to be observed. There can be little doubt that studies of the functional consequences of these interactions, in terms of mechanisms of secretion and of ticketing for specific destinations, will not only solve these major problems but will also contribute much to our understanding of the fine structure of membranes.

Genetic Studies on Protein Export in Bacteria

J. BECKWITH and S. FERRO-NOVICK[1]

1 Introduction

The study of protein export has proceeded in parallel in eukaryotic and prokaryotic systems over the past 10 years. Much of the information on eukaryotic protein secretion has come from the detailed analysis of an in vitro system which appears faithfully to mimic the in vivo pathway. In prokaryotes much of the progress has come from in vivo analysis including genetic studies. The genetic studies have focussed both on the components of specific proteins which determine their localization and on the nature of cellular components which are part of the export process. There now exists a large collection of mutations in the signal sequences of bacterial envelope proteins. These have been accumu-

[1] *Present address:* Department of Cell Biology, Yale University, School of Medicine, New Haven, CT 06510, USA

Department of Microbiology and Molecular Genetics, Harvard Medical School, Boston, MA 02115, USA

Current Topics in Microbiology and Immunology, Vol. 125

lated using both in vivo selections and in vitro constructions. These mutations have confirmed many aspects of the original signal sequence hypothesis and have provided some insight into the important features of such sequences. Studies on signal sequence mutations have been recently reviewed (MICHAELIS and BECKWITH 1982; BENSON et al. 1985) and will not be covered in this article.

Another approach to analyzing the role of the amino acid sequence of a protein in its export is the use of gene fusions. Such fusions have been constructed, again using both in vivo and in vitro techniques. These fusions code for hybrid proteins composed either of portions of two exported proteins or of portions of an exported protein and a cytoplasmic protein. The properties of gene fusion strains of this sort have provided information on several aspects of the export process. In addition, the phenotype of certain of these gene fusion strains has allowed genetic selections for mutants affecting secretion. The mutations thus obtained have either altered the export of specific proteins or appear to affect cellular components of a secretory apparatus. In addition to the use of gene fusions, other approaches have been employed to obtain mutants with pleiotropic effects on secretion. Studies on these mutants and characterization of the genes affected by them has led to new information on the protein export process in bacteria.

In this review we wish to summarize studies on the properties of gene fusions and cloned heterologous genes in bacteria which have shed light on the process of protein localization. In addition, we will describe the work done to this point on genes of *E. coli* which may be involved in the export process. Finally, in the process of summarizing these areas, we will refer occasionally to the similarities between the early steps in protein secretion in eukaryotic cells (transfer of proteins into the lumen of the rough endoplasmic reticulum) and the export of proteins across the cytoplasmic membrane of bacteria. Are these processes identical? Do they use the same general mechanism? Will genetic studies in bacteria contribute to an understanding of the secretory process in eukaryotic cells? These are questions that are still unresolved but merit a further analysis.

2 Gene Fusions and Recombinant DNA Clones in the Study of Protein Export

2.1 Introduction

The construction of gene or operon fusions has had wide utility in analyzing a variety of biological phenomena (SCHWARTZ 1985; SILHAVY and BECKWITH 1985; GUARENTE 1985). The protein export process is no exception. Gene fusions coding for hybrid proteins have been particularly useful in these studies. In this review, we will discuss those cases of gene fusion in which the amino-terminal portion of the hybrid protein includes the signal sequence from an exported protein. Gene fusions of several classes have been constructed – those which contain fusions of the signal sequence of one prokaryotic protein to the mature portion of another; those which contain the fusion of the signal sequence of a prokaryotic protein to the mature portion of a eukaryotic secreted protein; and those which contain the fusion of a signal sequence to a protein which is normally cytoplasmic. In conjunction with these studies, we will also discuss those cases where the gene for an exported protein from one organism has been cloned into a different organism. These will include eukaryotic genes cloned into bacteria and genes from one prokaryote cloned into another. The cellular location and form (e.g., precursor or processed) of the hybrid proteins or cloned gene products have been determined.

While the use of gene fusions and cloned genes has unquestionably answered a number of questions about the protein export process, there is one general limitation to these studies. The techniques available for determining the cellular location of a protein are fraught with difficulties and artifacts. For instance, different techniques for preparation of the periplasmic protein fraction yield somewhat different patterns of proteins (Beacham 1979). It is possible that one of the approaches causes leakage of some cytoplasmic proteins. Also, there can be problems in determining the cellular location of abnormal proteins. Such proteins include those coded for by gene fusions, fragments of proteins produced by deletion or chain-terminating mutants, and proteins coded for by genes from foreign organisms. In some cases, these proteins may form aggregates or have other properties which cause them to partition with membranes in various fractionation procedures, when their actual location is in one of the soluble compartments – the periplasm or the cytoplasm. The protein may then be mistakenly identified as a membrane protein. The case of an amber fragment of the periplasmic protein, β-lactamase, illustrates certain of these problems. Koshland and Botstein (1982) found that a procedure which releases the periplasmic fraction did not release this particular amino-terminal fragment of β-lactamase. The procedure disrupts the outer membrane of the bacteria, but leaves intact spheroplasts. The fragment was found in the spheroplast fraction. When the spheroplasts were broken, the fragment was released into the soluble fraction. The simplest interpretation of this result was that the amber fragment was cytoplasmic. However, subsequent studies on the accessibility of this fragment to proteolysis in intact spheroplasts suggested that, in fact, the protein was loosely adhering to the outer surface of the spheroplast (Koshland et al. 1982). These cautions should be kept in mind when reading this review of gene fusion studies and, in some cases, we will point out where such factors should be taken into account. What this brief summary of fractionation problems in bacteria points to is the need for improved or alternative techniques for accurately determining the location of proteins in bacteria.

2.2 Fusions Between Genes for Exported Prokaryotic Proteins

Certain major questions concerning protein export in bacteria can be approached by replacing the signal sequence of one prokaryotic protein with that of another. The first and simplest question is whether there is any match between the signal sequence and mature sequence of a particular protein; or can one exchange signal sequences between proteins at will without altering the ability of the cell to export a particular protein. This question has been answered in several studies. In one case, the signal sequence of the outer membrane porin PhoE was substituted with that of a similar protein OmpF, or vice versa (TOMMASSEN et al. 1983, 1984). In both cases, the proteins were localized in their active form to the outer membrane. Similarly, the signal sequence of the periplasmic TEM β-lactamase can be substituted for that of the periplasmic alkaline phosphatase and the latter enzyme is still found in the periplasm (HOFFMAN and WRIGHT 1985; MANOIL and BECKWITH 1985).

Are signal sequences merely signals for the initiation of the export process, or do they contain information for the ultimate location of a protein? For example, does an outer membrane protein require information in its signal sequence to reach its destination, or can a signal sequence for a periplasmic protein suffice? TOMMASSEN and coworkers found that when the PhoE protein had its signal sequence replaced with that of β-lactamase (TOMMASSEN et al. 1983), it still was properly exported to the outer membrane. This result indicates that the localization to the outer membrane involves amino acid sequences or structures within the mature portion of the PhoE protein. In a complementary experiment, it was shown that the signal sequence of the outer membrane protein OmpA can be substituted for that of β-lactamase, and the β-lactamase is still found in the periplasm (GHRAYEB et al. 1984). When more than the signal sequence of an outer membrane protein is added to a protein from another compartment, under some conditions the hybrid protein *will* be exported to the outer membrane. INOUYE and coworkers constructed a fusion in which the signal sequence plus nine amino acids of the mature portion of the outer membrane lipoprotein were fused to the normally periplasmic β-lactamase (GHRAYEB and INOUYE 1984). In this case, the β-lactamase was found in the outer membrane, presumably anchored by the amino-terminal sequence of lipoprotein which contains an acyl linkage to lipid.

Another issue of localization deals with certain inner membrane proteins. Many, if not most, proteins of the bacterial cytoplasmic membrane are not synthesized with cleavable amino-terminal signal sequences. These proteins include peripheral membrane proteins which are effectively cytoplasmic proteins bound to the membrane by their interaction with integral membrane proteins, and those which contain two or more hydrophobic membrane-spanning segments which anchor them tightly in the membrane. It is still not clear whether the incorporation of the latter class of protein into the membrane follows the same pathway as that of proteins with cleavable signal sequences. A third class of inner membrane proteins includes those with cleavable signal sequences which span the membrane only once. It has been presumed that these proteins are exported like periplasmic proteins, but remain anchored near their carboxy termini to the cytoplasmic membrane by that one segment. Evidence for this proposal comes from the studies of BOEKE and MODEL (1982), who have shown that deletion of the hydrophobic region of the protein results in a periplasmic location for the truncated polypeptide. Thus, the signal sequence for an integral (trans-membrane) protein can also function to export a protein to the periplasm.

Gene fusions have also been used to determine whether a protein designed for export in one organism can also cross the membrane barrier in another. Among bacteria, this question has been studied in several cases. The signal sequence from the α-amylase of the gram-positive *Bacillus amyloliquefaciens* has been fused to the mature portion of the TEM β-lactamase of gram-negative enteric bacteria (PALVA et al. 1982; ULMANEN et al. 1985). The α-amylase is normally secreted into the growth medium, while β-lactamase is normally periplasmic. When this gene fusion was introduced into *Bacillus subtilis*, 90% of the β-lactamase was found in the growth medium. Since there is no outer membrane in gram-positive bacteria and, therefore, no periplasmic space, it appears that β-lactamase can cross the cytoplasmic membrane of these bacteria. Similarly, the DNA for the signal sequence of the *E. coli* outer membrane protein, OmpA, was fused to the gene for the staphylococcal nuclease A of the gram-positive *S. aureus* and introduced into *E. coli* (TAKAHARA et al. 1985). Nearly all the nuclease activity was found in the supernatant after washing spheroplasts generated by osmotic shock. The authors propose that the enzyme is periplasmic, although some caution must be exercised in this conclusion, since the protein is not found in the original osmotic shock fluid.

In another case, the gene for the *Bacillus licheniformis* penicillinase, a major fraction of which is secreted into the medium in its native organism, was introduced into *E. coli*. The enzyme produced by the cloned genes was found bound to the *E. coli* outer membrane (LAI et al. 1981; HAYASHI and WU 1983; SARVAS and PALVA 1983). The reason for this location is presumably the existence of an amino acid sequence in penicillinase, quite similar to that found in the lipoprotein described above, which causes it to be modified and attached to lipid in the outer membrane.

These results together show that gram-positive signal sequences are recognized by the gram-negative bacteria as export signals and the protein is transported across the cytoplasmic membrane.

2.3 Expression of Eukaryotic Secreted Proteins in Bacteria

Evidence also exists that the gram-negative bacteria can recognize certain eukaryotic signal sequences and secrete and correctly process the proteins to which they are attached. In one case, the gene for rat preproinsulin was expressed in *E. coli* (TALMADGE et al. 1980a, b; CHAN et al. 1981). The protein was exported across the cytoplasmic membrane and correctly processed to yield periplasmic proinsulin. Similarly, the gene for human growth hormone was introduced into *Pseudomonas aeruginosa*, again yielding the correctly processed polypeptide in the periplasm (GRAY et al. 1984). Finally, the immunoglobulin light chain with its signal sequence coding region intact was cloned into *E. coli* and a portion of the immunoglobulin found in the periplasmic space (ZEMEL-DREASEN and ZAMIR 1984). The amount apparently exported was only around 25% of the total. In this paper, we note that the authors did not do controls to insure that the procedure for releasing periplasmic contents was not also resulting in some cell lysis. However, the fact that the protein found in the "periplasmic" fraction was processed supports their analysis.

2.4 Fusions of Prokaryotic Signal Sequences to Eukaryotic Proteins

Much of the work in this area has been done in the biotechnology companies in an effort to secrete in bacteria commercially useful proteins such as interferon, immunoglobulin, peptide hormones, and others. It is our impression from discussions with researchers in these companies that these efforts have met with mixed results. However, experiments in which a prokaryotic signal sequence fails to export a eukaryotic protein are probably not ordinarily reported in the literature. While it is clear from what is described below that it is certainly possible to obtain successful results from such experiments, they cannot be taken as a general rule. When the signal sequence of α-amylase was fused to the mature portion of human interferon a2 or mouse interferon B (PALVA et al. 1983; SHIROZA et al. 1985), the interferon was secreted and correctly processed in *Bacillus subtilis*. Similar results were obtained when the β-lactamase signal sequence was fused to proinsulin in *E. coli*. As with the intact eukaryotic gene, the hybrid gene coded for a protein which was exported to the periplasmic space (TALMADGE et al. 1980b). In the studies with immunoglobulin light chain, a construct was also introduced into *E. coli* in which the signal sequence plus a substantial portion of the mature β-lactamase protein was attached to the K light chain (ZEMEL-DREASEN and ZAMIR 1984). Again, about 20%–40% of the protein was found in the periplasm.

OHSUYE and coworkers (OHSUYE et al. 1983) have fused the signal sequence of alkaline phosphatase to the mature portion of neo-endorphin and introduced the construct into *E. coli*. They report that a portion of the hybrids were correctly processed, but remained cytoplasmic or on the inner surface of the cytoplasmic membrane. However, it seems possible to us that this conclusion may be based on the types of artifacts in cell fractionation we have discussed. All existing evidence suggests that the processing enzyme is on the external surface of the cytoplasmic membrane. If this is the case, processing is an indication of export. (We have assumed here that the processing seen is by the normal signal peptidase and not by some other cytoplasmic peptidase.) Examining this article in detail, we suspect that the hybrid proteins may be adhering to the external surface of the cytoplasmic membrane, leading to a mistaken interpretation of fractionation results.

We point out that in several cases the amounts of protein made from vectors carrying fused genes or heterologous cloned genes are less than expected. These lowered amounts could be due to various factors. Abnormal proteins or proteins from foreign organisms may be more sensitive to the proteases present in the cell. Such instability has been shown, for instance, for the gram-negative β-lactamase cloned into *Bacillus subtilis* (ULMANEN et al. 1985). Alternatively, there may be reduced efficiency of translation of the messenger RNAs for these proteins. Since in the eukaryotic secretion system it appears that translation and secretion are coupled, it is possible that foreign sequences or an inappropriate match between signal sequence and mature protein could potentiate a translational arrest mechanism. While there is evidence for proteolysis as the basis for reduced "expression," there is so far no support for the latter explanation. Careful studies on stability of these proteins are required (see Note Added in Proof).

All of these results suggest the following: (1) There is no necessary match between a signal sequence and the protein of which it is a part. That is, signal sequences can frequently be substituted one for another without any effect on the secretion process. (2) Eukaryotic proteins which pass through plasma membranes can often pass through the cytoplasmic membrane of bacteria. The interesting cases are those eukaryotic proteins which cannot be exported in bacteria. In the case of the immunoglobulin experiments, the export appears to be inefficient. (3) Gram-positive signal sequences are effectively recognized in gram-negative bacteria and vice versa. This was not necessarily an expected result, since the gram-positive signals are often considerably longer than the gram-negative ones. (4) Eukaryotic signal sequences are often recognized by the bacterial secretion system. This is perhaps not surprising, since the overall structure of eukaryotic and prokaryotic signal sequences and the nature of the processing site are quite similar. One difference which has been noted is that on the average eukaryotic signal sequences have one positive charge at their amino-termini while prokaryotic signal sequences have two (VON HEIJNE 1984).

These conclusions must be qualified because of the limited number of examples looked at, because of the presumed unreported failures, and because researchers have often not reported on the amounts of protein they are looking at.

Finally, gene fusions in both eukaryotic and prokaryotic systems have shed light on the question of whether signal sequences must be amino-terminal. TALMADGE et al. have constructed a fusion which codes for rat preproinsulin containing at its amino-terminus eight amino acids from β-galactosidase followed by ten amino acids coded for by DNA linkers (TALMADGE et al. 1981). Thus, the signal sequence (from rat preproinsulin) in this case is preceded by 18 "nonsignal" amino acids. Despite this tail, the bacteria are still able to export and process preproinsulin. In a eukaryotic system, the secretion of ovalbumin, which itself has an internal signal sequence located somewhere between positions 22 and 41 of the protein, was studied (TABE et al. 1984; KRIEG et al. 1984). This internal signal sequence is not processed. Further, it was found that adding on 21 amino acids to the amino-terminus of the protein did not prevent its transport into the endoplasmic reticulum of *Xenopus* oocytes. However, the addition of 51 amino acids, in the same way, did prevent the transfer process. Thus, clearly, with certain limitations, signal sequences need not be amino-terminal. It is curious, therefore, that there are not more cases of proteins with internal signal sequences.

2.5 Fusions of Signal Sequences to Cytoplasmic Proteins

A series of fusions have been isolated in which the amino-terminal region of the *E. coli* cytoplasmic protein β-galactosidase (coded for by the *lacZ* gene) has been replaced by amino-terminal sequences of the periplasmic proteins, maltose-binding protein (coded for by the *malE* gene) and alkaline phosphatase (BASSFORD et al. 1979; MICHAELIS et al. 1983a). In nearly all these constructs the intact signal sequence plus varying portions of the mature periplasmic protein are attached to β-galactosidase. In no case are the hybrid proteins exported to the periplasm. Instead, the proteins are incorporated into the cytoplasmic membrane. In this location, they appear to block the cell's secretory potential, causing the accumulation in the cell of precursors of most if not all exported proteins. It has been proposed that β-galactosidase has amino acid sequences within it which make it incompetent to pass through the membrane (VON HEIJNE 1980). In apparent contrast, fusions of β-galactosidase to amino-terminal regions of the outer membrane protein LamB (coded for by the *lamB* gene) have been reported to yield hybrid proteins which are exported to the outer membrane (SILHAVY et al. 1977). This apparent conflict may be explained in two ways. (1) The hybrid protein may be transported to the outer membrane not by passage through the periplasm, but rather by membrane-membrane contacts between the inner and outer membranes (SILHAVY et al. 1979). (2) Alternatively, the characterization of the hybrid proteins as outer membrane-bound may be an artifact of the type described above in our discussion of problems with cell fractionation. This has been suggested by TOMMASSEN and coworkers (TOMMASSEN et al. 1985).

Whatever the explanation, it is our feeling that most of the results with hybrid β-galactosidase molecules suggest that the β-galactosidase portion cannot pass through membranes. According to von Heijne's explanation, this would be due to the chance existence within the protein of sequences incompatible with passage through the membrane. Thus, we would not generalize from this one case to all cytoplasmic proteins. We would expect that some might be exported but others not, dependent on their amino acid sequence. A few examples are consistent with this view. On the one hand, the fusion of the β-lactamase signal sequence to the chicken cytoplasmic triose phosphate isomerase does not lead to export of the protein (KADONAGA et al. 1984). However, the fusion of the phage f1 coat protein sequence to the small lambda cro repressor protein yields a periplasmic hybrid protein (J.D. Boeke and P. Model, personal communication). In eukaryotic systems, the cytoplasmic β-globin can be sequestered in the lumen of the rough endoplasmic reticulum (TABE et al. 1984) of *Xenopus* oocytes when it is fused to the ovalbumin signal sequence or sequestered into microsomal vesicles when it is fused to the β-lactamase signal sequences (LINGAPPA et al. 1984). While these results are admittedly sparse, they are consistent with the views expressed above.

There are genetic approaches which should allow the characterization of those sequences such as in β-galactosidase which prevent export. If so, ultimately, they may allow the directed engineering of cytoplasmic proteins so that they can be exported across membranes.

2.6 Properties of Strains Containing Fusions of β-Galactosidase

2.6.1 Lethality of High Levels of Synthesis of Hybrid β-Galactosidase Molecules

A series of fusions have been isolated using in vivo techniques in which the amino-terminal regions of either maltose-binding protein (MBP) or bacteriophage λ receptor (LamB) are fused to β-galactosidase. The synthesis of these hybrid proteins is inducible by the addition of maltose to the growth medium, since both proteins play a role in maltose metabolism. As stated earlier, the hybrid proteins are localized to the cytoplasmic membrane. Detailed studies have been carried out on the effect on cells of the synthesis of MBP-β-galactosidase hybrids (ITO et al. 1981). As the concentration of the hybrid protein in the membrane increases after induction by maltose, there is an increasing inhibition of the export of cell envelope proteins. Precursors of these proteins appear in the cells and very little mature protein is found in its proper location. At early stages of induction, where the effect on export and processing is not so severe, the hybrid protein itself appears to undergo cleavage at its signal sequence (RASMUSSEN et al. 1984). However, by the time at which other proteins are found largely in precursor form, the same is true for the hybrid protein. Thus, it appears that the accumulation of the hybrid protein in the membrane is either directly blocking specific sites for protein export, or is so disrupting the membrane as to make further export of proteins impossible.

The cells are severely affected by the accumulation of the hybrid protein in the membrane. A substantial fraction of the cells eventually lyse. This again may be due either directly to a defect in secretion (e.g., some protein important for structure and stability may need to be exported) or due to disruption of the membrane. A similar phenomenon has been observed with hybrid proteins in which the amino-terminus of β-galactosidase has been replaced with that of OmpF (SODERGREN et al. 1985), LamB (EMR and SILHAVY 1982), alkaline phosphatase (MICHAELIS et al. 1983a), or PhoE (TOMMASSEN et al. 1985), although these systems have not been examined in as much detail.

Cells producing hybrid proteins between β-galactosidase and MBP or LamB exhibit a maltose-sensitive phenotype. Apparently, the block in protein export caused by the induction of these proteins is lethal to the bacteria. As a result, it is possible to select mutants of these strains which are resistant to the lethal effects of maltose. Certain of these mutants fail to incorporate the hybrid protein into the membrane, thus relieving the lethal effects. Most of these are due to mutations altering the signal sequence of the hybrid protein (EMR et al. 1978; BASSFORD and BECKWITH 1979). A second type of mutant obtained in this selection will be described below.

2.6.2 Altered Enzymatic Activity of Membrane-Bound β-Galactosidase

A second property of gene fusions of β-galactosidase to exported proteins has been quite useful. In the same cases described above, where a signal sequence is fused to β-galactosidase and the hybrid protein is membrane-bound, it appears that β-galactosidase is less able to assume its enzymatically active form. The specific way in which this is seen varies with the system. In the case of certain LamB-β-galactosidase hybrids, there is almost no measurable enzymatic activity even after full induction with maltose (HALL et al. 1982). In the case of one MBP-β-galactosidase hybrid, the uninduced levels of enzymatic activity are considerably lower than expected, while maltose induction results in more normal levels of activity (Bassford et al. 1979). In the case of alkaline phosphatase-β-galactosidase hybrids, there is measurable enzymatic activity both induced and uninduced, but it is lower than expected given the amount of protein present (Michaelis et al. 1983a).

This result is unexpected, since in a wide range of gene fusions that have been constructed with β-galactosidase it would appear that the enzymatic activity is independent of the nature of the amino-terminal end which is substituted for the normal amino-terminus. In fact, it seems likely that the low specific enzymatic activity of these membrane-bound hybrid proteins is not directly caused by the nature of the amino acid sequence at the amino-terminus, but rather is a result of the fact that this sequence leads the protein into the membrane. This is suggested by the finding that signal sequence mutants in the gene fusion result in a cytoplasmic location of the hybrid protein and restore a much higher specific activity to the protein.

In order to achieve an active conformation, identical monomeric subunits of β-galactosidase must assemble into tetramers. It may be that in the case of hybrid proteins which have been led into the membrane by signal sequences, the appropriate portions of the β-galactosidase moiety are not available for tetramerization. Alternatively, tetramers may form, but may not be active in the lipid environment of the membrane. However, the low enzymatic activity of membrane-bound β-galactosidase is not seen in all cases of membrane-bound enzyme. Certain hybrids of β-galactosidase with the integral membrane protein MalF exhibit more or less normal specific activities (Silhavy et al. 1976; Froshauer and Beckwith 1984; Froshauer 1985). We suspect that the difference may be explained as follows. The addition of a signal sequence of a periplasmic protein such as MBP (or of an outer membrane protein) to β-galactosidase results in the beginning of the export process. The signal sequence causes the MBP portion of the protein to be transferred through the membrane and then the β-galactosidase sequence follows. However, as suggested above, there may be sequences in β-galactosidase which prevent its passage through the membrane resulting in the imbedding of a portion of the latter protein in the membrane. According to this picture, a portion of β-galactosidase would still be protruding into the cytoplasm, but possibly not enough to allow tetramerization.

MalF is a protein which is thought to span the membrane many times. When β-galactosidase is fused to a position following certain of the presumed trans-membrane sequences, the hybrids with high specific activity are found. One can imagine that in these hybrid proteins the β-galactosidase portion is simply attached to the membrane by the trans-membrane segment of MalF and is itself entirely in the cytoplasm. The β-galactosidase would not be "imbedded" in the membrane as we suggest it might be in the case of the fusions to MBP and other periplasmic and outer membrane proteins.

At any rate, the enzymatic activity of hybrids of β-galactosidase with LamB, MBP, and alkaline phosphatase is such that strains producing them are unable to grow on lactose. As a result, selection for Lac^+ derivatives of these strains yields mutants which internalize the hybrid protein (Oliver and Beckwith 1981; Michaelis et al. 1983b; Emr and Silhavy 1982). These selections have generated, first of all, a collection of signal sequence mutants in these genes.

In addition, they have been used to obtain new classes of mutations with pleiotropic effects on protein export described below.

2.6.3 Cautions in the Use of lacZ Gene Fusions in the Study of Protein Localization

The properties of the *lacZ* gene fusions described here have provided both selections for mutants affecting protein export and have allowed various manipulations yielding new information on the export process in bacteria. However, there are clearly limitations to the degree to which the properties of these fusions and of mutants derived from them can be used to draw conclusions about the export process. One of the general approaches in using such fusions is to alter the portion of the hybrid protein derived from the exported protein either by selecting mutations or introducing alterations (e.g., deletions); the effect on the localization of the hybrid protein is then used to suggest the role of the altered sequences in export. However, we suggest that no solid conclusions on the role of such sequences can be drawn on the basis of the properties of the altered hybrid proteins alone. These are abnormal proteins. The combination of an exported protein with a cytoplasmic protein may generate a hybrid with unexpected properties.

A strong indication of the problems that might be encountered is seen in the case of certain signal sequence mutants isolated from a strain producing a LamB-β-galactosidase hybrid (Emr and Silhavy 1982). In particular, these mutations cause the substitution of charged amino acids for a glycine in the hydrophobic region of the LamB signal sequence. The altered proteins are now found predominantly localized to the cytoplasm instead of to the membrane. Based only on these characteristics, one might have concluded that this glycine residue was an important one for the functioning of the signal sequence. However, when these mutations were recombined from the *lamB-lacZ* fusion onto an *intact lamB* gene, no effect was seen on the export of LamB to the bacterial outer membrane. In other words, the β-galactosidase portion of the hybrid protein is not simply being passively carried by the LamB sequences; it can interact either directly or indirectly with the amino-terminal portion of the hybrid protein to alter its normal function. A misleading conclusion would have been drawn if the significance of these signal sequence mutants had been based only on their effects on localization of the hybrid protein.

The lesson from such an example is that studies with hybrid β-galactosidase molecules can only be used as indications of interesting features of the export process. If mutations have a certain effect on the properties of strains producing hybrid proteins, their effect on the intact gene for the exported protein must be checked. These qualifications would apply to various recent studies on *lamB-lacZ* and *malE-lacZ* fusion strains (Benson et al. 1984; Rasmussen et al. 1984). They would also apply to an earlier study on a *lamB-lacZ* fusion strain which led to the conclusion that a signal sequence did not contain sufficient information for protein export (Moreno et al. 1980). In that study, a hybrid protein containing the intact LamB signal sequence plus 15 amino acids of the mature LamB protein attached to β-galactosidase was found localized to the cytoplasm of *E. coli*. However, another explanation is still possible. The export process may have been initiated in this case, but the very short LamB sequence anchoring the protein in the cytoplasmic membrane may not have been sufficient to maintain the interaction with the membrane.

3 Pleiotropic Mutants Affecting Protein Export

3.1 Introduction

Eukaryotic cells have a complex secretory apparatus. The early steps of secretion, the passage of proteins through the membrane of the rough endoplasmic reticulum, appear to involve many proteins and at least one stable (7S) RNA molecule (Walter and Blobel 1980, 1982, 1983). These conclusions are based on in vitro studies. It is only recently that a comparable analysis has been initiated in bacteria (Muller and Blobel 1984; Chen et al. 1985). Results of Muller and Blobel indicate that there may also be a number of proteins required for the export process in bacteria. Several laboratories, including our own, have initiated genetic studies to identify components of a bacterial secretion apparatus. The general approach is to isolate mutants with pleiotropic effects on export, to use these mutants to identify important genes in the process and their products, and ultimately to correlate these in vivo studies with proteins identified in the in vitro system.

3.2 Selections for Pleiotropic Mutants Using Gene Fusions

Early selections used to obtain mutants with pleiotropic defects in protein export resulted in the isolation of *secA* (Oliver and Beckwith 1981) and *secB* (Kumamoto and Beckwith 1983) mutants. The selection for these mutants was based on the Lac^- properties of one of the MBP-β-galactosidase hybrid protein strains (see above, Sect. 2.6.2). Lac^+ derivatives of this strain were selected on lactose tetrazolium agar plates at 30° C. The Lac^+ colonies contained mutations in one of three loci: they were either in the signal sequence of the MBP-β-galactosidase hybrid, in the *secA* gene, or in the *secB* gene. All *secA* and *secB* mutants were partly defective for export of MBP at 30° C. Certain of the *secA*, but none of the *secB*, mutants, were conditional lethal; that is, they were unable to grow at 37° C or 42° C.

One of the *secAts* mutant strains, MM52, has been studied in detail. At high temperature, MM52 fails to export periplasmic or outer membrane proteins (Oliver and Beckwith 1981). So far, every protein from these compartments which has been examined is affected. While earlier results suggested that there may be a subset of periplasmic proteins which is unaffected, recent studies show that most, if not all of the major proteins found in this compartment are inhibited in their export (Liss and Oliver 1985). The proteins that fail to be exported accumulate in the cytoplasm in their precursor form. Furthermore, whereas in wild-type cells proteins such as MBP and alkaline phosphatase are synthesized on membrane-bound polysomes, in the *secAts* mutant they are made on free ribosomes (RASMUSSEN and BASSFORD 1985). Additional analysis has revealed that the *secA* gene product is a 92-kD protein which may be a peripheral cytoplasmic membrane protein (OLIVER and BECKWITH 1982a). The synthesis of this protein is derepressed about 10- to 20-fold in the *secAts* mutant. Its synthesis is increased in the same manner in other mutants which block protein export (*secY* and *secD*, see below). The *secA* gene maps at 2.5 min on the *E. coli* map at one end of a cluster of genes involved in cell envelope biogenesis and cell division (OLIVER and BECKWITH 1982b).

No conditional lethal mutants were found in the *secB* gene in the original selection. However, once the gene was mapped, it was possible to use various approaches to obtain null mutations, including insertions, in this gene. Surprisingly, the null *secB* mutants are not lethal; they can grow on minimal media but not on rich media (KUMAMOTO and BECKWITH 1985). The null mutants of *secB* block the export of MBP and OmpF, but have no effect on the export of ribose-binding protein or alkaline phosphatase. This specificity of the *secB* mutants raises the possibility that there may be different routes for certain steps in the export pathway in *E. coli.* The *secB* gene maps at 81 min on the *E. coli* map, directly adjacent to the structural gene for glycerol phosphate synthetase, *gpsA*.

More recent studies with the gene fusion technique have revealed that different classes of pleiotropic mutants can be isolated depending on the exported protein fused to β-galactosidase. Selections with strains harboring a *lamB-lacZ* gene fusion have yielded mutants in a gene termed *prlE*. The *prlE* mutants examined are cold-sensitive for growth and are blocked for the export of several proteins at the nonpermissive temperature (S.A. Benson and T. Silhavy, personal communication). With strains harboring a *phoA-lacZ* fusion, the *secD* mutant was isolated (Gardel, Hunt, Michaelis, and Beckwith, unpublished results). This mutant is also cold-sensitive and is blocked in the export of all proteins tested so far (MBP, LamB, PhoA, and OmpA). Both *secD* and *prlE* mutants map at 8.5 min on the *E. coli* chromosome.

The selections used to isolate *secA,B,D*, and *prlE* depended on the internalization and resultant increase in β-galactosidase activity of various hybrid proteins. The maltose-sensitive property (Sect. 2.6.1) of strains producing these hybrid proteins has also been used to isolate mutants affecting secretion. In particular, a *lamB-lacZ* fusion strain was employed. Selection for Mal^r derivatives yielded a mutant in the *prlF* gene, which maps at 70 min on the *E. coli* chromosome (KIINO and SILHAVY 1984). The *prlF* mutation reduces the growth of the bacteria at low temperatures and increases the efficiency at which proteins are exported from the cytoplasm. No effect on normal protein export has been observed. The isolation of null mutations should help to clarify the role of the *prlF* gene product.

3.3 Suppressors of Signal Sequence Mutants

Mutations in the *sec* genes were detected by their phenotype of causing a pleiotropic *defect* in secretion. Another class of pleiotropic mutants are those which result in a *restoration* of secretion of those proteins with mutationally defective signal sequences. One could imagine that certain classes of mutations altering components of the secretory machinery might allow improved export of such proteins. Mutations with these properties termed *prlA*,*B*, and *C* have been isolated as suppressors of LamB signal sequence mutants (Emr et al. 1981), while a mutation in the *prlD* gene was detected as a suppressor of an MBP signal sequence mutant (Bankaitis and Bassford 1985).

Mutations in the *prlA* gene can suppress the secretion defect in signal sequence mutants of LamB, MBP, and alkaline phosphatase (Emr and Bassford 1982; Michaelis et al. 1983b). In some cases, the suppression is highly efficient, with nearly complete restoration of export of the defective protein seen. The *prlA* gene maps at 72 min on the *E. coli* chromosome in an operon coding mainly for ribosomal proteins. However, it itself does not appear to be a ribosomal protein (Schultz et al. 1982; Shiba et al. 1984b). From the amino acid sequence of the protein as determined by DNA sequencing (Cerretti et al. 1983) and from fractionation properties of the PrlA protein (Ito 1984), it seems likely that it is located in the cytoplasmic membrane. Mutations in the *prlA* gene can restore secretion to proteins with severely defective signal sequences. These include a LamB mutant in which essentially the entire hydrophobic region of the signal sequence was removed (Emr et al. 1981) and a hybrid protein of alkaline phosphatase containing not even a remnant of a signal sequence (C. Hoffman and A. Wright, personal communication). In both cases, small but significant amounts of export of the defective proteins were observed.

At least two explanations come to mind for the *prlA* mutants. According to the first, a component of the normal secretory apparatus has been altered so as to allow the passage of proteins with defective signal sequences. According to the second, a new pathway has been opened up for protein export. Arguing against the second explanation are data which suggest that PrlA protein may be part of the normal pathway. Ito and coworkers detected mutations altering the expression of the *prlA* gene (*secY* mutations) which caused a pleiotropic defect in protein export (see below). The properties of these mutants suggest that PrlA is part of the normal export pathway. The fact that *prlA* mutants can restore some export to mutants with no or few facsimiles of signal sequences suggests that they may be exerting their effects by greatly reducing the specificity of the export pathway.

The *prlB* mutant weakly suppresses signal sequence mutations in *lamB*, but does not suppress mutations in *malE* (Emr and Bassford 1982). The export of LamB occurs without cleavage of the signal sequence. The *prlB* mutation has recently been shown to be a small deletion which removes a portion of the end of the gene for the periplasmic ribose-binding protein corresponding to the carboxy-terminus of the protein (Garwin and Emr, unpublished results). This bizarre mutant is still not understood. The *prlC* mutation can weakly suppress signal sequence mutation in *lamB* and *malE*. However, the processing of these proteins appears to be aberrant (Emr and Bassford 1982).

The *prlD* mutation was detected as a suppressor of a deletion which removes seven amino acids from the hydrophobic core of the MBP signal sequence (Bankaitis and Bassford 1985). The *prlD* mutation weakly suppresses some signal sequence mutants in *malE* and *lamB*, but does not suppress others. In fact, *prlD* enhances the export defect of certain *lamB* signal sequence mutants. The *prlD* gene maps at 2.5 min on the *E. coli* chromosome, but does not appear to be allelic to *secA*. When double mutants of certain *prlA* mutants and *prlD* are constructed, a partial block in export of MBP, RBP, LamB, and OmpA is observed. These results have been interpreted to mean that the *prlA* and *prlD* gene products form a complex which interacts with the signal sequence. Given the primitive level of understanding of the role of any of the *sec* or *prl* gene products in protein export, it seems premature to draw such conclusions.

3.4 Localized Mutagenesis

Another approach to seeking mutants with pleiotropic effects on protein export is to mutagenize those regions of the chromosome which include genes thought to play a role in the export process. Since it has been proposed that ribosomes may be intimately involved in secretion, it seemed possible that mutations in certain genes for ribosomal proteins might affect the export process.

Ito and his colleagues chose for such a study the region of the chromosome which includes the most ribosomal protein genes. This region lies close to min 72 on the *E. coli* map. Using a localized mutagenesis technique involving transduction with phage P1, they detected two mutations they termed *secY* mapping in this region which affected the export of maltose-binding protein (Ito et al. 1983; Shiba et al. 1984b). As indicated above, the mutations altered the secretion properties of the bacteria not because they were affecting a ribosomal protein, but rather because they affected the expression of the *prlA* gene. One of these mutations is an ochre mutation in a ribosomal protein gene upstream from *prlA*, which presumably exerts its effects by reducing the expression of *prlA* (Ito et al. 1984). The other is a mutation in the *prlA* structural gene (Shiba et al. 1984b). The *secY* mutants block the export of MBP and of the outer membrane proteins, OmpF and OmpA.

Suggestions that the *E. coli* 6S RNA might serve the same function as the 7S RNA of mammalian signal recognition particle led Lee and coworkers to mutagenize the structural gene for this stable RNA species (Lee et al. 1985). However, they showed that disruption of the gene, which eliminated 6S RNA from the cells, had no effect on protein export.

3.5 Different Mutations in the Same Gene Have a Different Phenotype

Until there is direct evidence for the role of the various *sec* and *prl* genes in the secretion apparatus, genetic and other in vivo studies can give an indication as to which genes are the best candidates. One indication that a gene may be part of the normal pathway in protein export is the finding of different classes of mutations in that gene which affect export in different ways. For instance, mutations in the *prlA* gene can suppress signal sequence mutations, block protein export, or suppress a *secAts* mutant (see below). In the *secAts* mutant, MM52, protein export is blocked and precursors accumulate in the cytoplasm. However, an amber mutant in *secA* has a different phenotype. In a strain containing a *secAam* mutation and a temperature-sensitive amber suppressor, little SecA protein is made at 42° C (Oliver and Beckwith 1982a). Depletion of the SecA protein does not simply prevent export of MBP and other proteins, it reduces their synthesis. This effect is not seen with all exported proteins; the synthesis of OmpA is unaffected (D.B. Oliver, personal communication). Initial results suggested that the block in MBP synthesis might be due to a coupling between secretion and translation (Kumamoto et al. 1984). However, more recent studies indicate that the effect on MBP synthesis is an indirect consequence of the loss of SecA protein from the cell; the block is reversed by the addition of 3′5′cyclic AMP to cells (Strauch and Beckwith, in preparation).

3.6 Suppressors of *secA* and *secY*

In order to identify more genes that could affect the export process in *E. coli*, suppressors of the *secA* and *secY* gene have been isolated (Ferro-Novick et al. 1984; Brickman et al. 1984; Oliver 1985; Shiba et al. 1984a). The identification of genes that can suppress the secretion defect in *secA* or *secY* could help us to understand the role that these gene products play in the process of protein export.

As a first step, the suppressor approach was used to see whether it would lead to the identification of genes that have already been implicated to play a role in secretion; Brickman et al. (1984) isolated suppressors of the *secAts* mutant (MM52) and then mapped these mutants. Two suppressors were linked to the previously identified *prlA* gene. One of these mutants (*sup1012*) was mapped more precisely and found to lie in or very close to the *prlA* gene. The *prlA* mutant that was isolated as a suppressor (*prlA1012*) of MM52 can suppress both growth defect and secretion defect of *secAts*. In the absence of the *secAts* mutation the *prlA1012* mutant grows slower than the wild-type mutant but has no effect on normal protein export. Unlike the *prlA* alleles previously isolated, the *prlA1012* mutation cannot suppress LamB signal sequence mutations but does weakly suppress a PhoA (*phoA61*) signal sequence mutation. Previously isolated *prlA* alleles that can suppress signal sequence mutations do not suppress MM52. The *prlA1012* mutant can suppress five different *secAts* alleles but cannot suppress the *secAam*.

The finding that *prlA1012* can suppress the *secAts* mutant implied that other interesting genes that play a role in secretion may be identified by using the suppressor approach. In order to facilitate the identification of additional genes that can mutate to suppress *secAts*, conditional lethal suppressors were isolated. The approach used was originally outlined by Jarvik and Botstein (1975). Spontaneous revertants of the *secAts* mutant were isolated and cold-sensitive growth mutants were screened for among the revertants (Ferro-Novick et al. 1984; Oliver 1985). This approach has led to the identification of the *secC* gene (FERRO-NOVICK et al. 1984) and the *ssaD*, *ssaE*, *ssaG*, and *ssaH* genes (OLIVER 1985).

The *secC* mutant can suppress the growth and secretion defect of *secAts*. In the absence of the *secAts* mutation, *secC* has a phenotype of its own. At 23° C (the cold-sensitive temperature), the *secCcs* mutant has a similar phenotype to the *secAam* mutant; it fails to synthesize certain exported proteins (MBP, RBP, LamB, and OmpF) (FERRO-NOVICK et al. 1984). Examination of whole cell lysates has revealed that, although the synthesis of MBP is completely blocked in *secCcs* at 23° C, a number of other proteins appear to be synthesized. For example, the cytoplasmic protein elongation factor-G still continues to be made. Like the *secAam* mutant, *secCcs* synthesizes the outer membrane protein OmpA at the restrictive temperature (S. Ferro-Novick and J.R. Beckwith, unpublished results). The *secCcs* mutant also synthesizes the periplasmic protein alkaline phosphatase at 23° C (Strauch and Beckwith, unpublished results). Therefore, it appears selectively to block the synthesis of certain proteins.

The map location of the *secC* gene (68.5 min of the *E. coli* chromosome) has aided in the identification of the *secC* gene product. The *secC* gene is identical to the *rpsO* gene which codes for the ribosomal protein S15, one of the proteins in the small subunit of the ribosome (Ferro-Novick et al., manuscript in preparation). The mechanism by which *secCcs* can suppress (*secAts*) remains to be determined.

Additional suppressors of the *secA* mutant, MM52, have been isolated and characterized by Oliver (OLIVER 1985). These mutants map at five new loci: *ssaD* (maps at 8.5 min on the *E. coli* chromosome); *ssaE* (50 min); *ssaF* (83 min); *ssaG* (41 min); and *ssaH* (91 min). All of the *ssa* mutants suppress both the growth and secretion defect of *secAts*. In the absence of the *secAts* mutation the *ssa* mutants are cs for growth and cause some reduction in the synthesis of MBP at the restrictive temperature. The *ssaF* gene product has recently been shown to code for the ribosomal protein L34 (*rpmH*), one of the proteins in the large subunit of the ribosome (D. Oliver, personal communication). The *ssaD* mutant, *secD*, and *prlE* all map in the same region of the chromosome (8.5 min). Genetic analysis is currently underway to determine whether these mutants are allelic.

Suppressors of the *secY* mutant have also been isolated. The approach used was essentially the same as with *secA*. The *ssyA3* mutant (SHIBA et al. 1984a) can suppress both the growth defect and secretion defect of *secY*. In addition, it is cold-sensitive for growth. At the cold temperature *ssyA3* reduces the rate of polypeptide chain elongation. At 42° C *ssyA3* suppresses the secretion defect of *secY* but has no effect on polypeptide chain elongation. ITO (1985) reports that newer studies have identified suppressor mutations in the gene for protein initiation factor 2 (*infB*) and in a gene which may, as with *secC*, be the *rpsO* gene.

Most recently, a new approach to isolating suppressors of a *secAts* mutation has revealed additional genes in which suppressor mutations may occur (C.A. Lee and J.R. Beckwith, unpublished results). These genes include ones coding for certain tRNA synthetases and for a tRNA! These various studies on suppressors raise the possibility that secretion defects can be overcome in some cases simply by an interference with protein synthesis. It is possible, then, that many of the suppressors described above act directly through their effects on protein synthesis.

3.7 Limitations and Problems in Characterizing Genes Important in the Secretion Process

While it is unclear how much of a secretion apparatus bacteria have, one can imagine several steps at which cellular proteins might be important in the process. By analogy with the eukaryotic system, these would be: (1) recognition of the signal sequence; (2) transport of the ribosome translating the protein to the membrane; (3) interaction with the membrane; (4) passage of the protein through the membrane; and (5) processing of the signal sequence. The selections which were used to obtain the *prl* mutants and the *sec* mutants probably require that the mutations affect steps relatively early in the process. The *prlA* mutants have altered the bacteria so that they can recognize defective signal sequences. The selections that yielded *secA*, *secB*, *secD*, and *prlE* demanded that the hybrid protein be internalized. It is unlikely that alterations or inactivation of the processing enzyme would show up in such selections. The evidence to date suggests that the export process is not obligatorily coupled to processing. It may be also that defects in any membrane apparatus that might exist for transporting proteins would not yield the phenotypes required in these selections. Furthermore, the *sec* selections require that the internalized hybrid protein be made. Therefore, defects in *sec* genes which result in a block in synthesis of these proteins would not show up.

Potentially, the characterization of extragenic suppressors might overcome these limitations. It is possible that suppressors in a gene later in the pathway might reverse the effects of a mutation such as *secAts*. In this way, one might "walk" down the pathway using suppressors. However, there are also limitations to the suppressor approach. Since, for ease of analysis, it is necessary to study only those suppressors which have a conditional lethal phenotype of their own, it is possible to miss genes of importance, either because such conditional lethal supressors cannot be isolated or because of the limitation of how many mutants can be screened.

Furthermore, studies with suppressors must be very carefully done, since there are many ways in which a particular mutant might be suppressed. Consider the mutation, *secAts*, presumably a missense mutation in the structural gene for a protein. The following classes of suppressors might be detected: (1) Missense suppressors which change a transfer RNA so that it can read the altered codon inserting back the original amino acid or an amino acid which will restore activity at the mutant site; (2) mutations in a protease might stabilize the altered protein product if it is unstable at high temperature; (3) mutations which affect cellular processes connected with the secretion process (e.g., protein synthesis) might compensate for the original defect. From the discussion of suppressors described above, it appears that this may be a major class of suppressors found for *sec* mutants; (4) mutations which alter some cellular component so as to substitute for the missing activity at the high temperature. These "bypass" mutations themselves are likely to be in genes related to the export process; and (5) mutations in components interacting with the mutant protein which restore a productive complex.

Thus, only careful studies will distinguish among these various classes of suppressors. First, the suppressor mutation should itself have a phenotype related to the export process, such as a pleiotropic defect in secretion at the nonpermissive temperature. If not, characterization of the gene and of mutants which eliminate its activity will be necessary to establish whether or not the gene plays a role in protein secretion.

3.8 Conclusions

We have described studies which reveal the existence of several genes of *E. coli* which may play an important role in secretion. At this point, we do not know if any of the genes code for proteins that are directly involved in export. However, those most likely to be involved are the *secA* and *prlA* genes. In the case of *prlA*, mutations with three different phenotypes affecting secretion have been characterized – signal sequence suppressors; those with pleiotropic defects in export; and suppressors of *secAts*. The properties of the *secAts* mutation plus its suppressibility by *prlA* mutants suggests that the *secA* gene product is also a component of the cell's apparatus for exporting proteins. Further, the derepression of synthesis of the SecA protein by various conditions which interfere with secretion is another indication that this gene may play a direct role in the export process. The *secB*, *secD* (*prlE*), *prlD*, and *prlF* genes have been less well-characterized, but are also candidates for genes of the secretion pathway. It is not clear whether the *rpsO* gene product (ribonsomal protein S15) is directly or indirectly involved in protein export. Further characterization of these latter genes should resolve some of these questions.

Ultimately, it is through the convergence of genetic studies and in vitro studies that the full description of the export process in bacteria will be achieved. Given the progress in these two areas, it would appear that that convergence is not far off.

Acknowledgments. We would like to thank Barry Bochner for help in doing a literature search and for comments on the manuscript. One of us (J.B.) did the writing of this manuscript while on sabbatical at the Department of Microbiology and Immunology, University of California, Berkeley. This work was supported by an American Cancer Society Research Professorship, by a National Science Foundation Grant to J.B., and by an American Cancer Society Postdoctoral Fellowship to S.F.N. We thank Ann McIntosh for excellent assistance in the preparation of this manuscript.

Note Added in Proof

A recent report (Matteucci and Lipetsky 1986) suggests that low expression of human growth hormone in *E. coli* is due to unexpected effects of messenger RNA sequence on gene expression. This may explain low expression of eukaryotic genes in *E. coli*, in general.

References

Bankaitis VA, Bassford PJ Jr (1985) Proper interaction between at least two components is required for efficient export of proteins to the *Escherichia coli*. J Bacteriol 161:169–178

Bassford P, Beckwith J (1979) Mutants of *Escherichia coli* which accumulate the precursor of a secreted protein in the cytoplasm. Nature 277:538–541

Bassford P, Silhavy T, Beckwith J (1979) Use of gene fusion to study secretion of maltose-binding protein in *Escherichia coli* periplasm. J Bacteriol 139:19–31

Beacham IR (1979) Periplasmic enzymes in gram-negative bacteria. Int J Biochem 10:877–883

Beckwith JR, Galizzi A, Smith GR (1985) The tools of bacterial genetics. In: Galizzi A, Scaife J (eds) Genetics of bacteria. Academic, London, pp 1–23

Benson SA, Bremer E, Silhavy TJ (1984) Intragenic region required for LamB export. Proc Natl Acad Sci USA 81:3830–3834

Benson SA, Hall MN, Silhavy TJ (1985) Genetic analysis of protein export in *Escherichia coli* K-12. Annu Rev Biochem 54:101–134

Boeke JD, Model P (1982) A prokaryotic membrane anchor sequence: carboxyl terminus of bacteriophage f1 gene III protein retains it in the membrane. Proc Natl Acad Sci USA 79:5200–5204

Brickman ER, Oliver DB, Garwin JL, Kumamoto C (1984) The use of extragenic suppressors to define genes involved in protein export in *Escherichia coli*. MGG 196:24–27

Cerretti DP, Dean D, Davis GR, Bedwell DM, Nomura M (1983) The spc ribosomal protein operon of *Escherichia coli*: sequence and cotranscription of the ribosomal protein genes and a protein export gene. Nucleic Acids Res 11:2599–2616

Chan SJ, Weiss J, Konrad M, White T, Bahl C, Yu S-D, Marks D, Steiner DF (1981) Biosynthesis and periplasmic segregation of human proinsulin in *Escherichia coli*. Proc Natl Acad Sci USA 78:5401–5405

Chen L, Rhoads D, Tai PC (1985) Alkaline phosphatase and OmpA protein can be translocated posttranslationally into membrane vesicles of *Escherichia coli*. J Bacteriol 161:973–980

Emr SD, Bassford PJ Jr (1982) Localization and processing of outer membrane and periplasmic proteins in *Escherichia coli* strains harboring export-specific suppressor mutations. J Biol Chem 257:5852–5860

Emr SD, Silhavy TJ (1982) The molecular components of the signal sequence that function in the initation of protein export. J Cell Biol 95:689–696

Emr SD, Schwartz M, Silhavy TJ (1978) Mutations altering the cellular localization of the phage λ receptor, an *Escherichia coli* outer membrane protein. Proc Natl Acad Sci USA 75:5802–5806

Emr SD, Hanley-Way S, Silhavy TJ (1981) Suppressor mutations that restore export of a protein with a defective signal sequence. Cell 23:79–88

Ferro-Novick S, Honma M, Beckwith J (1984) The product of gene secC is involved in the synthesis of exported proteins in *E. coli*. Cell 38:211–217

Froshauer S (1985) Structure, localization and regulation of MalF, an inner membrane protein for maltose transport in *Escherichia coli*. PhD Thesis, Harvard University

Froshauer S, Beckwith J (1984) The nucleotide sequence of the gene for malF protein, an inner membrane component of the maltose transport system of *Escherichia coli*. J Biol Chem 259:10896–10903

Ghrayeb J, Inouye M (1984) Nine amino acid residues at the NH_2-terminal of lipoprotein are sufficient for its modification, processing, and localization in the outer membrane of *Escherichia coli*. J Biol Chem 259:463–467

Ghrayeb J, Kimura H, Takahara M, Hsiung H, Masui Y, Inouye M (1984) Secretion cloning vectors in *Escherichia coli*. EMBO J 3:2437–2442

Gray GL, McKeown KA, Jones JSJ, Seeburg PH, Heyneker HL (1984) *Pseudomonas aeruginosa* secretes and correctly processes human growth hormone. Biotechnology Feb:161–165

Guarente L (1984) Use of gene fusions to study biological problems. In: Setlow J, Hollaender A (eds) Genetic engineering. Plenum, New York 6:233–251

Hall MN, Schwartz M, Silhavy TJ (1982) Sequence information within the lamB gene is required for proper routing of the bacteriophage λ receptor protein to the outer membrane of *Escherichia coli* K-12. J Mol Biol 156:93–112

Hayashi S, Wu HC (1983) Biosynthesis of *Bacillus licheniformis* penicillinase in *Escherichia coli* and *Bacillus subtilis*. J Bacteriol 156:773–777

Hoffman C, Wright A (1985) Fusions of secreted proteins to alkaline phosphatase: an approach for studying protein secretion. Proc Natl Acad Sci USA 82:5107–5111

Ito K (1984) Identification of the *secY* (*prlA*) gene product involved in protein export in *Escherichia coli*. MGG 197:204–208

Ito K (1986) Genetic control of protein secretion and localization. Adv Biophys 21

Ito K, Bassford PJ Jr, Beckwith J (1981) Protein localization in *E. coli*. Is there a common step in the secretion of periplasmic and outer membrane proteins? Cell 24:707–714

Ito K, Cerretti DP, Nashimoto H, Nomura M (1984) Characterization of an amber mutation in the structural gene for ribosomal protein L5, which impairs the expression of the protein export gene, secY, in *Escherichia coli*. EMBO J 3:2319–2324

Ito K, Wittekind M, Nomura M, Shiba K, Yura T, Miura A, Nashimoto H (1983) A temperature-sensitive mutant of *E. coli* exhibiting slow processing of exported proteins. Cell 32:789–797

Jarvik J, Botstein D (1975) Conditional-lethal mutations that suppress genetic defects in morphogenesis by altering structural proteins. Proc Natl Acad Sci USA 72:2738–2742

Kadonaga JT, Gautier AE, Straus DR, Charles AD, Edge MD, Knowles JR (1984) The role of the β-lactamase signal sequence in the secretion of proteins by *Escherichia coli*. J Biol Chem 259:2149–2154

Kiino DR, Silhavy TJ (1984) A mutation, *prlF1*, that relieves the lethality associated with export of β-galactosidase hybrid proteins in *Escherichia coli*. J Bacteriol 158:878–883

Koshland D, Botstein D (1982) Evidence for posttranslational translocation of β-lactamase across the bacterial inner membrane. Cell 30:893–902

Koshland D, Sauer RT, Botstein D (1982) Diverse effects of mutations in the signal sequence on the secretion of β-lactamase in *Salmonella typhimurium*. Cell 30:903–914

Krieg P, Strachan R, Wallis E, Tabe L, Colman A (1984) Efficient expression of cloned complementary DNAs for secretory proteins after injection into *Xenopus* oocytes. J Mol Biol 180:615–643

Kumamoto CA, Beckwith J (1983) Mutations in a new gene, secB, cause defective protein localization in *Escherichia coli*. J Bacteriol 154:253–260

Kumamoto C, Beckwith J (1985) Evidence for specificity at an early step in protein export in *Escherichia coli*. J Bacteriol 163:267–274

Kumamoto C, Oliver DB, Beckwith J (1984) Signal sequence mutations disrupt feedback between secretion of an exported protein and its synthesis in *E. coli*. Nature 308:863–864

Lai JS, Sarvas M, Brammer WJ, Neugebauer K, Wu HC (1981) *Bacillus licheniformis* penicillinase synthesized in *Escherichia coli* contains covalently linked fatty acid and glyceride. Proc Natl Acad Sci USA 78:3506–3510

Lee CA, Fournier MJ, Beckwith J (1985) *Escherichia coli* 6S RNA is not essential for growth or protein secretion. J Bacteriol 161:1156–1161

Lingappa VR, Chaidez J, Yost CS, Hedgpeth J (1984) Determinants for protein localization: β-lactamase signal sequence directs globin across microsomal membrane. Proc Natl Acad Sci USA 81:456–460

Liss LR, Oliver DB (1986) Effects of *secA* mutations on the synthesis and secretion of proteins in *Escherichia coli*: evidence for a major export system for cell envelope proteins. J Biol Chem 261:2299–2303

Manoil C, Beckwith J (1985) *TnPhoA*: a transposon probe for protein export signals. Proc Natl Acad Sci USA 82:8129–8133

Matteucci M, Lipetsky H (1986) Biotechnology 4:51–55

Michaelis S, Beckwith J (1982) Mechanism of incorporation of cell envelope proteins in *Escherichia coli*. Ann Rev Microbiol 36:435–465

Michaelis S, Guarente L, Beckwith J (1983a) In vitro construction and characterization of *phoA-lacZ* gene fusions in *Escherichia coli*. J Bacteriol 154:356–365

Michaelis S, Inouye H, Oliver D, Beckwith J (1983b) Mutations that alter the signal sequence of alkaline phosphatase of *Escherichia coli*. J Bacteriol 154:366–374

Moreno F, Fowler AV, Hall M, Silhavy TJ, Zabin I, Schwartz M (1980) A signal peptide is not sufficient to lead β-galactosidase out of the cytoplasm. Nature 286:356–359

Müller M, Blobel G (1984) Protein export in *Escherichia coli* requires a soluble activity. Proc Natl Acad Sci USA 81:7737–7741

Ohsuye K, Nomura M, Tanaka S, Kubota I, Nakazato H, Shinagawa H, Nakata A, Noguchi T (1983) Expression of chemically synthesized a-neo-endorphin gene fused to *E. coli* alkaline phosphatase. Nucleic Acids Res 11:1283–1294

Oliver DB (1985) Identification of five new essential genes involved in the synthesis of a secreted protein in *Escherichia coli.* J Bacteriol 161:285–291

Oliver D, Beckwith J (1981) *E. coli* mutant pleiotropically defective in the export of secreted proteins. Cell 25:765–772

Oliver D, Beckwith J (1982a) Regulation of a membrane component required for protein secretion in *Escherichia coli.* Cell 30:311–319

Oliver D, Beckwith J (1982b) Identification of a new gene (*secA*) and gene product involved in the secretion of envelope proteins in *Escherichia coli.* J Bacteriol 150:686–691

Palva I, Sarvas M, Lehtovaara P, Shakow M, Kaariainen L (1982) Secretion of *Escherichia coli* β-lactamase from *Bacillus subtilis* by the aid of α-amylase signal sequence. Proc Natl Acad Sci USA 79:5582–5586

Palva I, Lehtovaara P, Kaariainen L, Sibakov M, Cantell K, Schein CH, Kashiwagi K, Weissman C (1983) Secretion of interferon by *Bacillus subtilis.* Gene 22:229–235

Rasmussen BA, Bassford PJ Jr (1985) Both linked and unlinked mutations can alter the intracellular site of synthesis of exported proteins of *Escherichia coli.* J Bacteriol 161:258–264

Rasmussen BA, Bankaitis VA, Bassford PJ Jr (1984) Export and processing of malE-lacZ hybrid proteins in *Escherichia coli.* J Bacteriol 160:612–617

Sarvas MO, Palva IA (1983) The penicillinase of *Bacillus licheniformis* is an outer membrane protein in *Escherichia coli.* J Bacteriol 155:657–663

Schultz J, Silhavy TJ, Berman ML, Fiil N, Emr SC (1982) A previously unidentified gene in the *spc* operon of *Escherichia coli* K12 specifies a component of the protein export machinery. Cell 31:227–235

Schwartz M (1985) Gene fusions in bacteria. In: Scaife J, Galizzi A Genetics of bacteria. Academic, London, pp 65–84

Shiba K, Ito K, Yura T (1984a) Mutation that suppresses the protein export defect of the *secY* mutation and causes cold-sensitive growth of *Escherichia coli.* J Bacteriol 160:696–701

Shiba K, Ito K, Yura T, Cerretti DP (1984b) A defined mutation in the protein export gene within the spc ribosomal protein operon of *Escherichia coli*: isolation and characterization of a new temperature-sensitive *secY* mutant. EMBO J 3:631–635

Shiroza T, Nakazawa K, Tashiro K, Yamane K, Yanagi K, Yamasaki M, Tamura G, Saito H, Karvade Y, Taniguchi T (1985) Synthesis and secretion of active mouse interferon-B using a *Bacillus subtilis* α-amylase secretion vector. Gene 34:1–8

Silhavy TJ, Beckwith JR (1985) Uses of lac fusions for the study of biological problems. Microbiol Rev 43:398–418

Silhavy TJ, Casadaban MJ, Shuman HA, Beckwith JR (1976) Conversion of β-galactosidase to a membrane-bound state by gene fusion. Proc Natl Acad Sci USA 73:3423–3427

Silhavy TJ, Shuman HA, Beckwith J, Schwartz M (1977) Use of gene fusions to study outer membrane protein localization in *Escherichia coli.* Proc Natl Acad Sci USA 74:5411–5415

Silhavy TJ, Bassford PJ, Beckwith J (1979) A genetic approach to the study of protein localization in *E. coli.* In: Inouye M (ed) Bacterial outer membrane: biosynthesis, assembly and functions. Wiley, New York, pp 203–254

Sodergren EJ, Davidson J, Taylor RK, Silhavy TJ (1985) Selection for mutants altered in the expression or export of outer membrane porin OmpF. J Bacteriol 162:1047–1053

Tabe L, Krieg P, Strachan R, Jackson D, Wallis E, Colman A (1984) Segregation of mutant ovalbumins and ovalbumin-globin fusion proteins in *Xenopus* oocytes. J Mol Biol 180:645–66

Takahara M, Hibler DW, Barr PJ, Inouye M (1985) The ompA signal peptide directed secretion of staphyloccoccal nuclease A by *Escherichia coli.* J Biol Chem 260:2670–2674

Talmadge K, Kaufman J, Gilbert W (1980a) Bacteria mature preproinsulin to proinsulin. Proc Natl Acad Sci USA 77:3988–3992

Talmadge K, Stahl S, Gilbert W (1980b) Eukaryotic signal sequence transports insulin antigen in *Escherichia coli.* Proc Natl Acad Sci USA 77:3369–3373

Talmadge K, Brosius J, Gilbert W (1981) An internal signal sequence directs secretion and processing of proinsulin in bacteria. Nature 294:176–178

Tommassen J, van Tol H, Lugtenberg B (1983) The ultimate locatization of an OM protein of *Escherichia coli* K12 is not determined by the signal sequence. EMBO J 2:1275–1279

Tommassen J, Pugsley AP, Korteland J, Verbakel J, Lugtenberg B (1984) Gene encoding a hybrid OmpF-PhoE pore protein in the outer membrane of *Escherichia coli* K12. MGG 197:503–508

Tommassen J, Leunissen J, van Damme-Jongsten M, Overduin P (1985) Failure of *E. coli* K-12 to transport PhoE-LacZ hybrid protein out of the cytoplasm. EMBO J 4:1041–1047
Ulmanen I, Lundstron K, Lehtovaara P, Sarvas M, Ruohonen M, Palva I (1985) Transcription and translation of foreign genes in *Bacillus subtilis* by the aid of a secretion vector. J Bacteriol 162:176–182
von Heijne G (1980) Trans-membrane translocation of proteins: a detailed physico-chemical analysis. Eur J Biochem 103:431–438
von Heijne G (1984) Analysis of the distribution of charged residues in the N-terminal region of signal sequences: implications for protein export in prokaryotic and eukaryotic cells. EMBO J 3:2315–2318
Walter P, Blobel G (1980) Purification of a membrane-associated protein complex required for protein translocation across the endoplasmic reticulum. Proc Natl Acad Sci USA 77:7112–7116
Walter P, Blobel G (1982) Signal recognition particle contains a 7S RNA essential for protein translocation across the endoplasmic reticulum. Nature 299:691–698
Walter P, Blobel G (1983) Disassembly and reconstitution of signal recognition particle. Cell 34:525–533
Zemel-Dreasen O, Zamir A (1984) Secretion and processing of an immunoglobulin light chain in *Escherichia coli*. Gene 27:315–322

The Membrane Trigger Hypothesis Revisited

B. WICKNER

A cardinal rule of modern scientific life, one which I have shamelessly violated, is that one should not write too many reviews. In this spirit, and fearing that Gunther Blobel and I were invited to spar in print here merely to promote book sales, I at first declined the kind invitation from the editors. However, they threatened to nominate me for service on an NIH study section. Thus, to minimize pain, I take pen in hand. What follows is not intended as a scholarly review, but rather opinions of the successes, and failures, of the membrane trigger hypothesis.

Through no fault of its proponents (and, rather, due to the excellence of their writing and experiments), the signal hypothesis took the scientific world by storm. By 1978, undergraduates and graduate students had no difficulty in describing how secreted polypeptides emerge from a ribosome, thread through a protein pore in the membrane, and emerge into the lumen of the endoplasmic reticulum (ER). I have reviewed the virtues, and limitations, of this model previously (WICKNER 1979, 1980). Another school of thought, grounded in the work of BRETSCHER (1973) and TANFORD (1973), supposed that considerations of physical chemistry and of protein folding might be important for protein insertion into, and across, membranes. Thus, in the space of a few years, von Heijne introduced the direct transfer model, which first presented a quantitative treatment of the energetics of protein transfer through a bilayer. Inouye presented the loop model, which pointed out that adjacent regions of a polypeptide chain could help each other through a bilayer. These ideas were fused in the helical hairpin hypothesis of Engelman and Steitz.

The membrane trigger hypothesis, which I proposed in 1979 (WICKNER 1979), suggested that proteins emerge from the ribosome into an essentially aqueous environment and will fold in response to that environment, that is, with apolar residues in the center of the protein and polar residues exposed on the protein surface. During transit into the hydrocarbon core of a lipid bilayer, the protein is likely to refold, and may again change its conformation when it emerges on the other side of the membrane. A second element in this scheme is that the cleaved, amino-terminal leader sequence found on so many membrane and secreted proteins emerges first from the ribosome and may influence the folding pathway of the subsequent residues. This seems especially likely in light of the fact that protein domains often fold in the millisecond time scale, whereas the synthesis of a protein takes many seconds or even a few

Department of Biological Chemistry, University of California, Los Angeles, CA90024, USA

Current Topics in Microbiology and Immunology, Vol. 125

minutes. These considerations led to a few predictions: (1) Proteins would pass through the membrane in direct contact with the lipid, rather than through a polar, proteinaceous core. (2) Proteins should be in a different conformation prior to inserting through the membrane than when in the membrane. (3) Secretion might be either cotranslational or post-translational, and mediated by a lipid or proteinaceous receptor, but the ribosome itself would have no *direct* role in the secretion process. (4) Some of the "information" for export should be in the mature sequence, and the match of the leader and mature should be important.

Clearly, both my writing and experimentation left something to be desired. Far from taking the scientific world by storm, this hypothesis caused a bit of a storm! However, to paraphrase W.C. Fields, reports of its demise (WALTER and BLOBEL 1981) were exaggerated. Several observations made during the past 6 years are in good agreement with self-assembly models of protein assembly into membranes: (1) Protein assembly into mitochondria or export across the bacterial plasma membrane are clearly not coupled to polypeptide chain growth and, in fact, are largely post-translational. There are even examples of post-translational protein import into the lumen of the ER. However, it is clear that most proteins which enter that organelle do so during their synthesis. (2) Bacterial exported proteins have been found in different conformations prior to their assembly; whether these particular conformations are essential to export or merely coincidental remains to be determined. (3) Bacterial leader peptides have been found to interact directly with amphiphiles, and genetic studies suggest that this interaction is vital to protein export. (4) In bacteria, a leader peptide alone is insufficient to cause translocation of the remainder of the protein across the membrane. However, leader peptides *are* sufficient to direct even β-galactosidase into the lumen of the ER or into mitochondria in eukaryotic cells.

More important than extolling the successes of the membrane trigger hypothesis is clearly to state some of its failures. Insufficient emphasis was placed on the role of receptor proteins in protein targeting. In eukaryotes, the concept of separately decoded packets of assembly information present in different parts of the protein has been remarkably fruitful. However, a recent consideration of the assembly of multispanning integral membrane proteins of the ER suggests that folding of a protein may be essential for insertion, that is, that there are spontaneous insertion "domains" in a folded nascent chain as well as insertion sequences.

Each hypothesis of the 1970s has been useful to stimulate experimentation and even debate. None was entirely true or untrue. All are showing signs of age. Harvey Lodish and I have recently attempted to salvage scraps from each of these relics and launch a "consensus" version (WICKNER and LODISH 1985); this too will undoubtedly need revision. More important, the coming years hold prospects for progress on the following fronts:

1. Reconstitution of translocation of preproteins across bilayers of fully defined protein composition and topology. So far, this goal has only been met with M13 procoat protein.

2. Isolation of any proteins needed for translocation.
3. Determination of the role of the electrical potential in those translocation systems where it is needed.
4. Determination of the important features of the leader sequence for each step (binding, translocation, and covalent processing) of assembly through the membrane.

Clearly, the combination of molecular genetics and enzymology, especially in microorganisms, holds promise of elucidating basic mechanisms in the near future.

References

Bretscher MS (1973) Science 181:622
Tanford C (1973) The hydrophobic effect. Wiley, New York
Walter P, Blobel G (1981) J Cell Biol 91:551
Wickner W (1979) Annu Rev Biochem 48:23
Wickner W (1980) Science 210:861
Wickner W, Lodish H, Science 230:400

In Vitro Analysis of the Bacterial Protein Export

M. MÜLLER[1] and G. BLOBEL[2]

1 Introduction

Based on the results of in vitro studies the signal hypothesis has been introduced (BLOBEL and DOBBERSTEIN 1975) as a model to describe sorting and subsequent transport of newly synthesized proteins across and into, respectively, the membrane of the endoplasmic reticulum. Some of the predictions of the signal hypothesis have recently been verified by the identification and biochemical characterization of two components of the transport machinery of the endoplasmic reticulum. One of them is called the signal recognition particle (SRP), a peripheral membrane ribonucleoprotein complex that specifically recognizes the nascent polypeptide chains of secretory and integral membrane proteins (WALTER et al. 1981). The other component is the SRP receptor (GILMORE et al. 1982), a single polypeptide which is integrated into the membrane of the endoplasmic reticulum. It is also called docking protein (MEYER et al. 1982).

In an in vitro translation-translocation system SRP – by binding to the ribosome – arrests the elongation specifically of secretory and integral membrane proteins. The block is released only when microsomes are added to which the ribosome attaches via SRP and its receptor so that translocation can be initiated (WALTER and BLOBEL 1981). Essential features of the protein transport into the endoplasmic reticulum, therefore, are (1) the temporal coupling of polypeptide chain elongation and transport (cotranslational mode) and (2) the involvement of proteinaceous translocation factors (receptor-mediated process).

[1] Biochemisches Institut der Universität Freiburg, D-7800 Freiburg
[2] Laboratory of Cell Biology, The Rockefeller University, New York, NY 10021, USA

Current Topics in Microbiology and Immunology, Vol. 125

Protein export in gram-negative bacteria comprises translocation across as well as integration into the inner and outer membrane of the cell envelope. Protein transport across the bacterial plasma membrane shares characteristic features with that of the endoplasmic reticulum: (1) Like most eukaryotic secretory and integral membrane proteins, bacterial proteins that are exported are synthesized with transient signal sequences. (2) Eukaryotic secretory proteins are correctly processed and localized in vivo by the transport machinery of the bacterial plasma membrane (refs. listed in Müller and Blobel 1984a). (3) A prokaryotic secretory protein requires SRP for segregation within the endoplasmic reticulum (Müller et al. 1982).

On the other hand, various conflicting results have been interpreted in the past as being contradictory to the notion of a common mechanism of protein transport across both membranes. Thus, both a co- and a post-translational mode of translocation were reported to be operative in bacteria. The major coat protein of filamentous phages which is transiently integrated into the bacterial plasma membrane prior to the virus assembly was found to integrate into lipid bilayers without the assistance of export factors (Watts et al. 1981).

The isolation of mutants of *Escherichia coli* that are pleiotropically defective in the export of proteins, however, provided strong evidence for the existence of bacterial gene products generally involved in protein export (Ito et al. 1983; Michaelis and Beckwith 1982; Silhavy et al. 1983).

In the following article we describe experiments which were devised to analyze biochemically components of the bacterial transport machinery. The results are discussed in view of the above-mentioned comparison between eukaryotic and prokaryotic protein transport systems.

2 Bacterial Protein Export Can Be Reproduced in an In Vitro System

The basic strategy of our analysis was to set up first an in vitro system that would allow a faithful reproduction of the in vivo translocation of exported proteins across the plasma membrane of *E. coli*. An ideal such system ought to synthesize efficiently selected proteins and translocate them into the lumen of added plasma membrane vesicles. A protein-synthesizing extract was obtained by preparing a high-speed supernatant of *E. coli* MZ 9 cells that was devoid of endogenous DNA, mRNA, and membranes. Synthesis selectively of exported proteins was achieved by programming the cell-free extract with plasmids that contained the genes of exported proteins under the control of sufficiently strong promotors. The system could be optimized such that the radioactively labeled newly synthesized proteins after separation on sodium dodecyl sulfate polyacrylamide gel electrophoresis (SDS-PAGE) could be visualized by autoradiography after a few hours of exposure only.

Translocation into cotranslationally added plasma membrane vesicles from *E. coli* that have an inside-out orientation (inverted vesicles, INV; Futai 1974) was followed by digesting soluble proteins with proteinase K. Since the protease

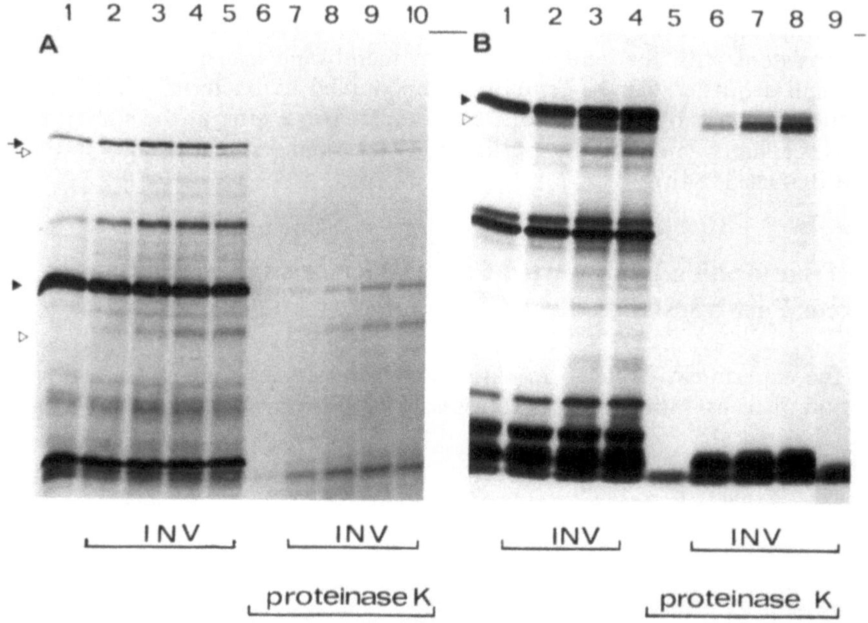

Fig. 1 A, B. In vitro translation and translocation into INV of alkaline phosphatase, truncated β-lactamase, and LamB protein. Plasmid pHI-1 (**A**) and plasmid pLB 7012B (**B**) were used to program a transcription-translation system prepared from *E. coli* strain MZ9. Translocation products were separated by polyacrylamide gel electrophoresis in SDS and visualized by fluorography. Translations were either in the absence or, where indicated, in the presence of INV at concentrations of 0.16, 0.4, 0.8, and 1.2 A_{280} units/ml (**A**, lanes 2–5 and lanes 7–10, respectively) and of 0.15, 0.5, and 1.0 A_{280} units/ml (**B**, lanes 2–4 and 6–8, respectively). Post-translational proteolysis with proteinase K was carried out (where indicated) at 22° C and for 30 min at a concentration of 200 µg proteinase K/ml. In a control, proteinase K was added together with Triton X-100, final concentration of 1% (wt/vol) (**B**, lane 9). *Closed symbols* indicate precursors and *open symbols* indicate mature forms of alkaline phosphatase (**A**, *arrow*), truncated β-lactamase (**A**, *arrowhead*), and LamB protein (**B**, *arrowhead*)

does not have access to proteins located within closed membrane vesicles any labeled protease-resistant protein must have been translocated into the INV. Figure 1 illustrates the results obtained with two periplasmic and one outer membrane protein. Plasmid pHI1 (obtained from H. Inouye) directed the synthesis predominantly of the precursors of alkaline phosphatase and of a truncated β-lactamase (panel A, closed arrow, and arrowhead, respectively) whereas plasmid pLB7012 B (obtained from S. Benson) gave rise to the precursor of the LamB protein (panel B, closed arrowhead). All three precursors were partially converted into the respective mature forms when INV were present during the translation (A, lanes 2–5; B, lanes 2–4, open symbols with INV added in increasing amounts). The mature forms of all three proteins were resistant toward proteinase K (A, lanes 7–10; B, lanes 5–8) unless the integrity of the vesicles was disrupted by detergent (B, lane 9).

By the criteria outlined above those protease-resistant molecules therefore must have been translocated into the INV. The same applies for a minor portion of each precursor (A, lanes 7–10; B, lanes 6–8, closed symbols), a finding which is consistent with the general idea that membrane transport and cleavage of a signal sequence can be temporally uncoupled in bacteria. That conversion of the precursor by INV was due to a correct processing at the signal sequence cleavage site was shown by partial NH_2-terminal sequence analysis (Müller and Blobel 1984a).

3 Translocation into Inverted Plasma Membrane Vesicles Can Occur Post-translationally

In the experiment depicted in Fig. 1 inverted vesicles were present during the period of translation. In order to assess whether translocation of proteins could

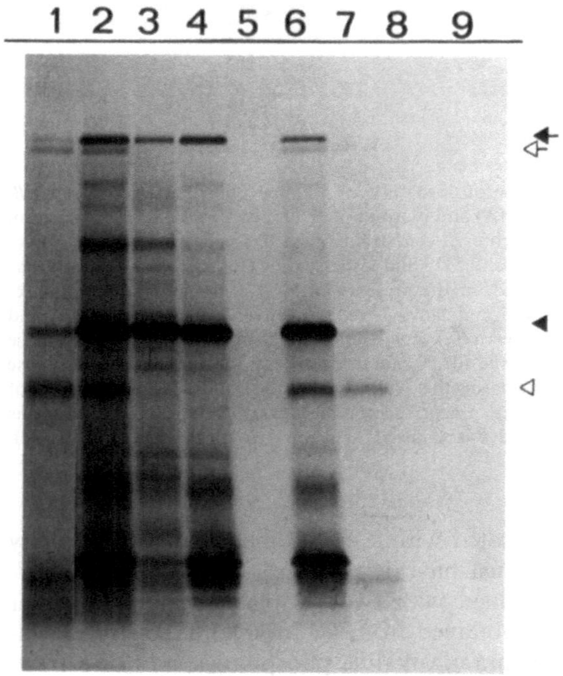

Fig. 2. In vitro translocation of periplasmic proteins can occur post-translationally. pHI-1-directed translation products are shown. Experimental details and symbols are as described in the legend to Fig. 1. Products synthesized during 30 min in the presence of INV (0.5 A_{280} unit/ml) are shown before (*lane 2*) or after (*lane 1*) digestion with proteinase K (200 μg/ml). Also shown are products synthesized in the absence of INV and either sedimented (*lane 3*) or remaining in the supernatant (*lane 4*) after centrifugation through a cushion of 0.5 *M* sucrose. The "supernatant" products were incubated for 30 min at 37° C with an equal volume of translation buffer containing INV (1.0 A_{280} unit/ml) (*lane 6*). Supernatant products digested with proteinase K (200 μg/ml) are shown before (*lane 5*) or after (*lane 7*) incubation with INV. A control is shown in *lane 8*: translation was carried out in the presence of 0.4 μ*M* nonradioactive methionine and in the absence of INV; after the centrifugation step there was an additional incubation of the supernatant with [^{35}S]-methionine. *Lane 9*, as lane 8 except for addition of ribosomes

occur uncoupled from their synthesis, INV were added to completed precursor chains of alkaline phosphatase and truncated β-lactamase (Fig. 2) under conditions that would strictly prohibit elongation of previously initiated peptides. This precaution was taken in order to avoid a putative release of a SRP-mediated translation arrest by the addition of membrane vesicles. Such a membrane-induced elongation would lead to labeled proteins segregated within the vesicles which then would erroneously be scored as post-translationally translocated molecules. Therefore, precursors were freed from all monosomes and polysomes possibly containing arrested peptides and were subsequently incubated with INV. As can be seen in Fig. 2, lanes 6 and 7, this resulted in the post-translational translocation of both precursors into the vesicles. A post-translational mode of translocation has been demonstrated in previous reports using similar in vitro systems (CHEN et al. 1985; DECKER et al. 1982; GOODMAN et al. 1981; OXENDER et al. 1980), none of them, however, clearly ruling out an INV-mediated release of prelabeled arrested peptides. Whether the export of bacterial proteins in vivo is obligatorily post-translational is not addressed by the experimental approach described here nor by the numerous reports of other authors emphasizing the post-translational mode in bacteria (CROWLESMITH et al. 1981; INOUYE et al. 1982; ITO et al. 1979; JOSEFSSON and RANDALL 1981; KOSHLAND and BOTSTEIN 1982; RANDALL 1983).

4 In Vitro Translocation into Plasma Membrane Vesicles is Abolished upon Subfractionation of the Cell-Free System

On the premise that protein export in bacteria requires distinct factors we tried to demonstrate the existence of such factors by modifying our in vitro system in a way that would render translocation into INV dependent on particular subcellular fractions. Assuming that a putative export factor might be soluble in the cytoplasm, in analogy to the eukaryotic SRP, we decided to subfractionate the nonparticulate part of our in vitro system.

The strategy of subfractionation was based on a quantitative reduction of the number of soluble proteins to a certain minimum that would still allow protein synthesis to go on that, however, would not translocate the newly synthesized proteins into added plasma membrane vesicles. In brief, protein synthesis was reconstituted from salt-extracted ribosomes, purified *E. coli* initiation factors 1, 2, 3, and a particular subfraction of a postribosomal supernatant that contained elongation factors, release factors, and tRNA synthetases. The latter was inferred from the fact that it complemented translation activity of ribosomes and initiation factors. This reconstituted system could be operated as a translation system and – when supplemented with RNA polymerase – also as a simultaneous transcription-translation system. Some proteins were synthesized as effectively as in the complete system (Fig. 3, lower closed arrow); others like alkaline phosphatase or the lambda receptor could only be expressed if their genes were put under the control of strong, inducer- and repressor-independent promoters (data not shown). The most outstanding characteristic of our reconstituted sys-

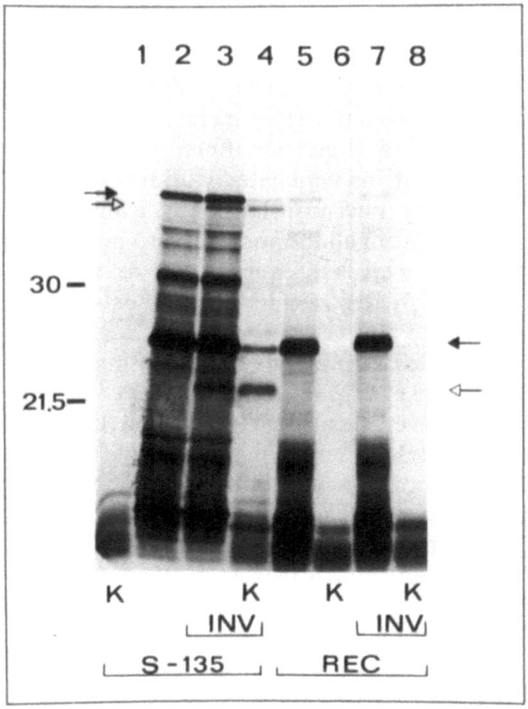

Fig. 3. In vitro translocation into INV of truncated β-lactamase is lost upon subfractionation of the in vitro system. For experimental details see legend to Fig. 1. *K*, proteinase K-resistant proteins in *lanes 1, 4, 6*, and *8*. The corresponding total translation products are shown in *lanes 2, 3, 5*, and *7*, respectively. *S-135*, complete in vitro system; *REC*, reconstituted in vitro system. *Rightward-pointing arrows*, alkaline phosphatase; *leftward-pointing arrows*, truncated β-lactamase. *Numbers to the left* indicate molecular mass in kilodaltons of standard proteins

tem, however, was that it had lost most of its translocation activity into the nonmodified inverted plasma membrane vesicles (cf. lanes 4 and 8).

5 In Vitro Translocation is Restored in the Reconstituted System by a Distinct Export Factor

The finding that translocation activity was drastically reduced after subfractionation of our in vitro system prompted us to use the transport-inactive system for assaying and thereby identifying bacterial export factors. Again in analogy to the eukaryotic SRP, which is an 11S particle, we shought to isolate bacterial counterparts by a possibly high sedimentation velocity that would discriminate them from the bulk of soluble proteins. In the experiment shown in Fig. 4 the reconstituted system was supplemented with 19 fractions each obtained after separating the soluble proteins (postribosomal supernatant) on a sucrose gradient. In Fig. 4B are shown only translocated proteins after post-translational digestion of nonmembrane enclosed proteins by proteinase K. In this experiment the radioactive counts of each sample that was run on the gel had been normalized so that only changes in the rate of translocation rather than in that of synthesis would be detected. Under these conditions an increase in the background translocation of our reconstituted system (panel B, control)

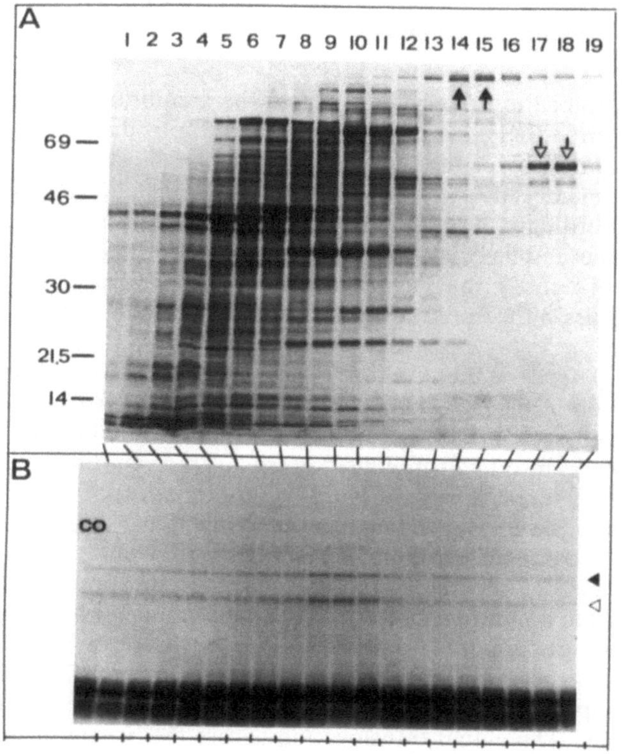

Fig. 4A, B. Sucrose gradient fractionation of a postribosomal supernatant. **A** A 10-μl aliquot of each of the 19 sucrose gradient fractions was analyzed by electrophoresis in 10%–15% polyacrylamide gradient gels in SDS. Shown is the Coomassie blue staining pattern. The β, β' subunits of RNA polymerase in fractions 14 and 15 are indicated by *upward-pointing arrows*. The subunit of the groE polymer in fractions 17 and 18 is indicated by *downward-pointing arrows*. *Numbers to the left* indicate molecular mass in kilodaltons of standard proteins. **B** Six-microliter aliquots of sucrose gradient fractions 1–19 were analyzed for their ability to affect translocation in the reconstituted system containing pHI-1 plasmid DNA, salt-washed ribosomes, IF1, IF2, and IF3, sucrose gradient fraction 5 or 6, and INV. In a control (*co*), buffer was added instead of a gradient fraction. The proteinase K-protected (i.e., translocated) polypeptides were analyzed by polyacrylamide gel electrophoresis in SDS and subsequent fluorography. The *closed arrowhead* points to the precursor form and the *open arrowhead* points to the mature form of truncated β-lactamase

was observed with fractions of a small range located to the middle of the gradient (lanes 9–11). This stimulation was exclusively due to an enhanced rate of translocation into INV since protein synthesis proved to be unaffected by those fractions (data not shown).

The translocation-stimulating activity could be enriched by adsorption to and high-salt elution from a diethylaminoethanol (DEAE) column and by precipitation with ammonium sulfate (Müller and Blobel 1984b). It was effective in mediating transport of several exported proteins like β-lactamase, alkaline phosphatase, and the lambda receptor. From its sedimentation behavior on the sucrose gradient (Fig. 4) a tentative sedimentation coefficient of 12S could be calculated. This might indicate that the export factor is an oligomeric protein.

6 Concluding Remarks

Translocation of proteins across the membranes of the endoplasmic reticulum requires SRP. As pointed out in the introductory section, SRP functions in binding polypeptide chains during their biosynthesis to the transport sites of the membrane, i.e., it coordinates synthesis and transport of secretory proteins. Accordingly, a cotranslational mode of translocation of proteins across membranes implies that the process is factor (receptor) mediated. The reverse conclusion, namely that a post-translational transport of proteins does not require the assistance of transport factors is not consistent with the results presented here.

Therefore the question arises, how the post-translational fashion of translocation in bacteria can be reconciled with the requirement for an export factor. A theoretical interpretation of our data is based on a dissection of the transport process into two subsequent events: (1) recognition of the precursor, (2) translocation across the membrane. The bacterial export factor described in here might – like the SRP – function in recognizing a nascent bacterial precursor leading to a complex between the nascent polypeptide chain and the factor. Unlike in the SRP-mediated process, the bacterial export factor might not immediately bind to a putative receptor in the membrane nor arrest translation, thus allowing a completion of the precursor prior to transport. In such a situation the recognition process, leading to a sorting of exported from nonexported proteins by complexation with the export factor, would be the essential cotranslational event. The actual translocation across the plasma membrane could be a subsequent process that, however, would require complexed precursors primed for transport rather than single precursor molecules.

The completion of nascent chains prior to translocation is not consistent with an SRP-like translation arrest. Although such arrest has never been demonstrated to be operative in bacteria a temporal coupling between synthesis and transport has repeatedly been reported following genetic analysis of the export process. It is conceivable that this regulation of synthesis involves transcription rather than translation.

References

Blobel G, Dobberstein B (1975) Transfer of proteins across membranes. J Cell Biol 67:835–851

Chang CN, Model P, Blobel G (1979) Membrane biogenesis: Cotranslational integration of the bacteriophage f1 coat protein into an Escherichia coli membrane fraction. Proc Natl Acad Sci USA 76:1251–1255

Chen L, Rhoads D, Tai PC (1985) Alkaline phosphatase and OmpA protein can be translocated posttranslationally into membrane vesicles of Escherichia coli. J Bacteriol 161:973–980

Crowlesmith I, Gamon K, Henning U (1981) Precursor proteins are intermediates in vivo in the synthesis of two major outer membrane proteins, the OmpA and OmpF proteins, of Escherichia coli K12. Eur J Biochem 113:375–380

Decker KP, Brusilow WSA, Gunsalus RP, Simoni RD (1982) In vitro membrane association of the F_0 polypeptides of the Escherichia coli proton translocating ATPase. J Bacteriol 152:815–821

Futai M (1974) Orientation of membrane vesicles from Escherichia coli prepared by different procedures. J Membr Biol 15:15–28
Gilmore R, Walter P, Blobel G (1982) Protein translocation across the endoplasmic reticulum. II. Isolation and characterization of the signal recognition particle receptor. J Cell Biol 95:470–477
Goodman JM, Watts C, Wickner W (1981) Membrane assembly: Posttranslational insertion of M13 procoat protein into E. coli membranes and its proteolytic conversion to coat protein in vitro. Cell 24:437–441
Inouye S, Soberon X, Franceschini T, Nakamura K, Itakura K, Inouye M (1982) Role of positive charge on the amino-terminal region of the signal peptide in protein secretion across the membrane. Proc Natl Acad Sci USA 79:3438–3441
Ito K, Mandel G, Wickner W (1979) Soluble precursor of an integral membrane protein: Synthesis of procoat protein in Escherichia coli infected with bacteriophage M13. Proc Natl Acad Sci USA 76:1199–1203
Ito K, Wittekind M, Nomura M, Shiba K, Yura T, Miura A, Nashimoto H (1983) A temperature-sensitive mutant of E. coli exhibiting slow processing of exported proteins. Cell 32:789–797
Josefsson LG, Randall LL (1981) Different exported proteins in E. coli show differences in the temporal mode of processing in vivo. Cell 25:151–157
Koshland D, Botstein D (1982) Evidence for posttranslational translocation of β-lactamase across the bacterial inner membrane. Cell 30:893–902
Meyer DI, Krause E, Dobberstein B (1982) Secretory protein translocation across membranes – the role of the "docking protein". Nature 297:647–650
Michaelis S, Beckwith J (1982) Mechanism of incorporation of cell envelope proteins in Escherichia coli. Annu Rev Microbiol 36:435–464
Müller M, Blobel G (1984a) In vitro translocation of bacterial proteins across the plasma membrane of Escherichia coli. Proc Natl Acad Sci USA 81:7421–7425
Müller M, Blobel G (1984b) Protein export in Escherichia coli requires a soluble activity. Proc Natl Acad Sci USA 81:7737–7741
Müller M, Ibrahimi I, Chang CN, Walter P, Blobel G (1982) A bacterial secretory protein requires signal recognition particle for translocation across mammalian endoplasmic reticulum. J Biol Chem 257:11860–11863
Oxender DL, Anderson JJ, Daniels CJ, Landick R, Gunsalus RP, Zurawski G, Yanofsky C (1980) Amino-terminal sequence and processing of the precursor of the leucine-specific binding protein, and evidence for conformational differences between the precursor and the mature form. Proc Natl Acad Sci USA 77:2005–2009
Randall LL (1983) Translocation of domains of nascent periplasmic proteins across the cytoplasmic membrane is independent of elongation. Cell 33:231–240
Silhavy TJ, Benson SA, Emr SD (1983) Mechanisms of protein localization. Microbiol Rev 47:313–344
Walter P, Blobel G (1981) Translocation of proteins across the endoplasmic reticulum. III. Signal recognition protein (SRP) causes signal sequence-dependent and site-specific arrest of chain elongation that is released by microsomal membranes. J Cell Biol 91:557–561
Walter P, Ibrahimi I, Blobel G (1981) Translocation of proteins across the endoplasmic reticulum. I. Signal recognition protein (SRP) binds to in-vitro-assembled polysomes synthesizing secretory protein. J Cell Biol 91:545–550
Watts C, Silver P, Wickner W (1981) Membrane assembly from purified components. II. Assembly of M13 procoat into liposomes reconstituted with purified leader peptidase. Cell 25:347–353

Biochemical Studies of Bacterial Protein Export*

P.C. TAI[1]

1 Introduction

Studies on the mechanisms of protein secretion and transport have been, until recently, confined to animal cells. In the past few years genetic and biochemical studies with bacteria have contributed greatly to our knowledge of mechanisms of protein secretion and localization (reviews by DAVIS and TAI 1980; WICKNER 1980; OSBORN and WU 1980; MICHAELIS and BECKWITH 1982; RANDALL and HARDY 1984; BENSON et al. 1985). In particular, selection of mutants has permitted the definition of the function of the cleavable signal sequence and the identification of many gene products that may be involved in protein export (see Beckwith and Ferro-Novick, this volume; Lunn et al., this volume). In addition, several signal peptidases have been purified and characterized and their genes

* Dedicated to Bernard D. Davis on the occasion of his 70th birthday

[1] Department of Metabolic Regulation, Boston Biomedical Research Institute, Boston, MA 02114, USA and Department of Microbiology and Molecular Genetics, Harvard Medical School, Boston, MA 02115, USA

Current Topics in Microbiology and Immunology, Vol. 125

identified (DATE and WICKNER 1981; INNIS et al. 1984; YU et al. 1984; ZWIZINSKI and WICKNER 1980; RAY et al., this volume). Moreover, since the outer surface of the bacterial membrane is readily accessible to manipulation, we have been able to use extracellular radioactive labeling or protease digestion of protruding secreted nascent chains to demonstrate directly the cotranslational secretion (SMITH et al. 1977, 1978) that had been inferred from the earlier in vitro work. The study of accessibility of nascent chains to protease has also revealed that a large domain of secreted nascent chains accumulates on the inner surface of the cytoplasmic membrane before the chains reach the outer surface (RANDALL 1983).

This article will focus on two areas in which our work provides further biochemical analysis of the mechanisms of protein export in bacteria: (1) protein translocation into *Escherichia coli* membrane vesicles and (2) interaction of ribosomes and membranes of *Bacillus subtilis*. The discussion will be highly selective and is not intended to cover the complete literature.

2 Protein Translocation into E. coli Membrane Vesicles

An efficient in vitro translocation system would be of considerable value in the further analysis of the molecular mechanisms of protein transport, especially in conjunction with many mutants altered in the process. Despite several reported successes (CHANG et al. 1979; SMITH 1980; DANIELS et al. 1980; GOODMAN et al. 1981), a reliable bacterial system has until recently been elusive. It has thus not been possible to determine whether protein translocation across bacterial membranes resembles that across eukaryotic membranes in requiring a signal recognition particle for translational arrest (WALTER and BLOBEL 1980, 1981) and for interaction with the receptor-docking protein in the membrane (GILMORE et al. 1982; MEYER et al. 1982).

Recently we developed a system for translocating bacterial proteins into a selected fraction of inverted *E. coli* cytoplasmic membrane vesicles (RHOADS et al. 1984; CHEN et al. 1985). It was found that a normally cotranslationally translocated protein can also be translocated post-translationally. The ability to separate translocation from translation and the high efficiency of the in vitro system, despite its intrinsic limitations, provide a tool for the detailed analysis of translocation.

2.1 Reconstituted Protein Translocation System

In developing a reliable system for translocating *E. coli* alkaline phosphatase and OmpA protein, synthesized in vitro from enriched mRNA, into *E. coli* inverted membrane vesicles, we found that the lighter subfractions of the cytoplasmic membranes are more active than the total cytoplasmic membranes (RHOADS et al. 1984; CHEN et al. 1985). Translocation is measured by protection of the products from digestion by pronase and by their sedimentation with the membrane vesicles and is, usually but not always (see Sect. 2.3), accompanied by cleavage of the signal sequence from a precursor. In an apparently cotranslational translocation system, in which membrane vesicles are added during protein synthesis, compounds that affect membrane fluidity or protonmotive force (but see Sect. 2.4) inhibit the translocation (RHOADS et al. 1984), as was observed by others in intact bacterial cells. Treatment of the vesicles with protease, under conditions that did not interfere with subsequent protein synthesis, also inactivates them for subsequent translocation. It is concluded that translocation of certain proteins across cytoplasmic membranes requires energy and membrane protein component(s) (RHOADS et al. 1984).

Polyamines (spermidine and putrescine) have been found to improve greatly the consistency and the amount of translocation of alkaline phosphatase and OmpA protein (CHEN et al. 1985). In the presence of these agents, the system is capable of carrying out not only cotranslational but also post-translational translocation, both proteins being translocated when membrane vesicles are added after cessation of protein synthesis. Under appropriate conditions post-translational translocation is as efficient as cotranslational translocation and also requires energy and protein-containing components in the membrane (CHEN et al. 1985). Another in vitro protein translocation system, capable also of post-translational translocation, has recently been described (MULLER and BLOBEL 1984a; this volume). Post-translational translocation of bacterial proteins in intact cells has been previously reported (JOSEFSSON and RANDALL 1981; KOSHLAND and BOTSTEIN 1982; GARWIN and BECKWITH 1982; Beckwith and Ferro-Novick, this volume).

2.2 E. coli Can Translocate the Same Protein Co- and Post-translationally

Although post-translational translocation would be expected for OmpA protein, which is synthesized on free polysomes (Table 1; CHEN et al. 1985), it is unexpected for alkaline phosphatase, which in cells is made predominantly on membrane-bound polysomes (CHEN et al. 1985; VARENNE et al. 1978). Indeed, the cotranslational translocation of alkaline phosphatase in vivo has been directly demonstrated by extracellular labeling of nascent peptide chains growing on membrane-bound polysomes (SMITH et al. 1977). The post-translational translocation of alkaline phosphatase in vitro thus indicates that it can be translocated either co- or post-translationally. Another secreted enzyme, R-TEM beta-lactamase, and another outer membrane protein, matrix protein, are made by both free and membrane-bound polysomes in *E. coli* cells (Table 1; CHEN et al. 1985). Moreover, many precursors that accumulate in export-deficient mutants can be chased, albeit slowly, to the proper location (Beckwith and Ferro-Novick, this volume). It therefore appears that *E. coli* has the flexibility of translocating the same proteins either cotranslationally or post-translationally, perhaps depending on how quickly the signal peptide of the nascent chain reaches the membrane (Fig. 1).

Table 1. Site of Synthesis of Proteins in *E. coli*

Protein	% Synthesis by		
	Membrane Polysomes	Free Polysomes	Protein Location
OmpA	<10	>90	OM
Matrix	66	34	OM
APase	86	14	periplasm
β-Lactamase			
w/enrichment	>90	<10	periplasm
w/o enrichment	46	54	
EFTu, EFG	<10	>90	cytoplasm

Polysomes from *E. coli* strain CW3747 were run off and immunoprecipitated for OmpA, matrix protein, and alkaline phosphatase. Polysomes prepared from strain C600/pBR322 with or without L broth enrichment were assayed for β-lactamase synthesis. EFTu or EFG were also assayed in each of the preparations; results were identical. The fluorograms were scanned and analyzed (from Chen et al. 1985)

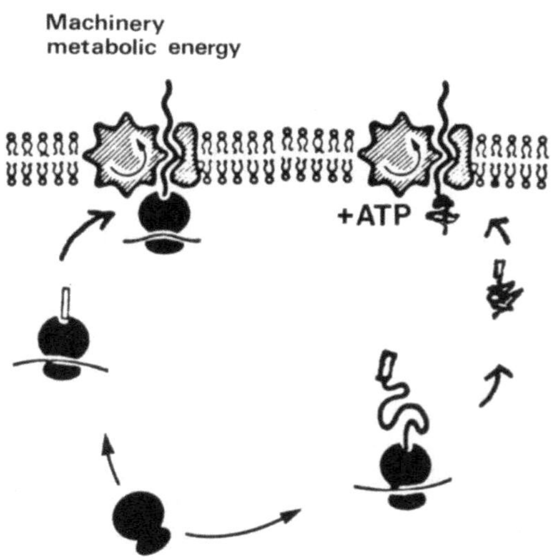

Fig. 1. Proposed model for protein translocation in *E. coli*. The same protein could be translocated co- or post-translationally, depending on how quickly the signal peptide of the nascent chain reaches the membrane. Both processes would utilize the protein-bearing membrane machinery, and ATP as energy source. An ATP-binding site and lipid-protein interaction in the membrane are involved in the process. A cytoplasmic factor(s) may also participate

2.2.1 Lack of Translational Arrest in the In Vitro System

In contrast to this flexibility, translation and secretion are reported to be tightly coupled in eukaryotic systems: a signal recognition particle causes translational arrest (Walter and Blobel 1980, 1981) and is released on subsequent interaction with a receptor in the (dog pancreatic) membrane (Meyer et al. 1982; Gilmore et al. 1982). Although genetic evidence suggests a coupling of synthesis and secretion in *E. coli* cells (Oliver and Beckwith 1981; Benson et al. 1985; Beckwith and Ferro-Novick, this volume), so far we have not been able to detect an analogous translational arrest. Indeed, such tight coupling may be a special property of the wheat germ extract and dog pancreatic membrane and its signal recognition particle, since the translational arrest and its relief by the membrane were also observed with bacterial β-lactamase in that system (MULLER et al. 1982). Whether bacterial membranes require the function of an analogous signal recognition particle in interacting with precursors and receptor protein remains to be determined. In this regard, a protein complex that appears to be involved in secretion has been detected in *B. subtilis* (CAULFIELD et al. 1984; see also Sect. 3) and in *E. coli*. An export factor required in a translocation system also appears to be oligomeric (MÜLLER and BLOBEL 1984b).

2.2.2 Competency of the Precursor for Translocation

Earlier studies had shown that the translocation of alkaline phosphatase into *E. coli* membrane vesicles occurs exclusively cotranslationally (SMITH 1980). The failure to observe post-translational translocation is explained by our finding that the ability of the precursor to be translocated decays in a time-dependent manner after completion of protein synthesis (CHEN et al. 1985). Similarly, the inactivation of the precursor of phage coat protein by Mg^{2+} accounts for the apparent lack of its post-translational insertion into membrane vesicles (GOODMAN et al. 1981; CHANG et al. 1979). It may well be that a specific conformation of a precursor molecule is required in order for it to be competent for translocation. It follows that improper folding of a precursor (or its signal peptide), either due to inactivation or mutation, may render it incompetent for translocation. The state of competency of a precursor may explain why precursors of phosphate-binding protein, accumulated in the cytoplasm of hyperproducing *E. coli* cells, are not exported post-translationally, while some precursors associated with membrane are exported (PAGES et al. 1984).

2.3 Cleavage of Signal Peptide Can Be Dissociated from Translocation

We have originally defined the cleavage of the signal peptide from a precursor as one of the criteria for successful translocation in our in vitro system. However, an appreciable fraction (10%–20%) is evidently translocated in the precursor form, since molecules of precursor size are found to be resistant to protease treatment and to cosediment with the membrane (RHOADS et al. 1984; CHEN et al. 1985). Moreover, the protection of precursor from proteolysis is temperature and energy dependent, just as the mature protein (see below). The translocation without cleavage of a signal peptide is in accord with the earlier findings that a mutant prolipoprotein and mutant precursors of LamB protein (LIN et al. 1978; SILHAVY et al. 1983) and β-lactamase precursor (KOSHLAND and BOTSTEIN 1982) are translocated in bacterial cells. In addition, we have found that *E. coli* prolipoprotein can be translocated in vitro without cleavage of the signal peptide. These results indicate that normal precursors can be translocated without the removal of the signal peptide by a signal peptidase (see also

MÜLLER and BLOBEL 1984a). Conversely, precursors can be processed even when translocation is blocked by inhibitors or by inactivation of the membrane vesicles by protease (RHOADS et al. 1984; CHEN et al. 1985; see also ZIMMERMANN and WICKNER 1983). Accordingly, processing of the precursors is neither a prerequisite nor necessarily a consequence of translocation and thus not a reliable indicator of either cotranslational or post-translational translocation. Furthermore, the results described above suggest that sites of signal peptide cleavage in the membrane are distinct from those of protein translocation.

2.4 ATP is Essential for Protein Translocation

It has been concluded that protonmotive force is required for protein translocation: inhibitors of protonmotive force prevent the cleavage and maturation of many precursors in *E. coli* cells (BAKKER and RANDALL 1984; DANIELS et al. 1981; ENEQUIST et al. 1981; JOSEFSSON and RANDALL 1981; PAGES and LAZDUNSKI 1982; ZIMMERMANN and WICKNER 1983). However, this conclusion may be questioned in view of the fact that the cleavage of a precursor is not a reliable indicator of translocation (Sect. 2.3). This question could be approached with the in vitro system. Since in vitro translocation can be accomplished post-translationally, the energy requirement for translocation could be examined after removal of components necessary for protein synthesis. Under such conditions, we found that translocation of both alkaline phosphatase and OmpA protein into *E. coli* membrane vesicles occurs in the presence of ATP or an ATP-generating system, or, less efficiently, in the presence of D-lactate oxidation (CHEN and TAI 1985). D-Lactate-driven translocation may be due to ATP synthesized via H^+-ATPase by the protonmotive force, the translocation being prevented by inhibitors of respiration or proton uncouplers as well as by hexokinase (plus glucose) or *E. coli* glycerol kinase (plus glycerol). The latter is highly specific for ATP as substrate and should have no effect on protonmotive force. This conclusion is supported by the finding that ATP can support protein translocation (though less efficiently) even in the presence of proton uncouplers that inhibit D-lactate-driven translocation. Moreover, membrane vesicles of mutants defective in the F_1-portion of ATPase, which cannot hydrolyze ATP to generate protonmotive force nor utilize D-lactate, can still use ATP to translocate proteins (CHEN and TAI 1985). It is thus evident that ATP, not protonmotive force, is essential for the post-translational translocation of alkaline phosphatase and OmpA protein into inverted *E. coli* membrane vesicles. Since *E. coli* can translocate the same protein either co- or post-translationally, ATP is probably also necessary for cotranslational translocation in bacteria. (However, such a requirement would be difficult to demonstrate because of the involvement of ATP in protein synthesis.)

2.4.1 *Roles of Protonmotive Force and H^+-ATPase*

Although protonmotive force is not essential for translocation, it contributes to the process, ATP-driven translocation being reduced 50%–70% in the presence of a proton uncoupler (CHEN and TAI 1985). One role of protonmotive force may be to maintain overall topology and organization of membrane proteins that are involved in translocation of the precursor, ensuring its proper interaction with signal peptidase for cleavage of the signal peptide and for the release of the mature protein. In addition, protonmotive force may contribute to the faster kinetics of translocation and processing of precursors in the bacterial cells.

Another role of protonmotive force may be to replenish ATP via H^+-ATPase in the in vitro system, as is clearly the case in the D-lactate-driven translocation (CHEN and TAI 1985). Furthermore, in our earlier translocation system in which translocation and translation were not uncoupled, protein translocation (but not synthesis of presursors) was prevented by inhibitors of protonmotive force (RHOADS et al. 1984). It is now clear that protonmotive force was required for the regeneration of ATP in that system, since excess phosphoenolpyruvate to regenerate ATP relieved the inhibition (unpublished). Depletion of ATP could also explain the puzzling observation that early addition of membrane vesicles during protein synthesis, or the presence of an excessive amount of membrane, not only failed to enhance but actually decreased translocation (RHOADS et al. 1984; CHEN et al. 1985). Thus H^+-ATPase seems to play a dual role in the in vitro translocation system: on the one hand it contributes to deplete ATP, on the other it regenerates ATP through protonmotive force.

In the coupled transcription-translation translocation system of MÜLLER and BLOBEL (1984a, b), a functional F_1 portion of H^+-ATPase is reported to be indispensable. This contrasts with our observation that membrane vesicles prepared from mutant defective in this portion could still translocate proteins with ATP (Sect. 2.4). It is possible that the requirement for F_1 of H^+-ATPase might be in part for regeneration of ATP which could be depleted severely in the coupled system. A direct role of H^+-ATPase in protein translocation can be ruled out by our recent findings that membrane vesicles prepared from a mutant in which whole *unc* operon has been deleted is capable of carrying out ATP-driven protein translocation (unpublished).

2.4.2 ATP Hydrolysis and Nucleotide Specificity

The requirement of ATP for translocation apparently involves its hydrolysis, since several nonhydrolyzable analogs of ATP (AMP-PNP, AMP-PCP, or AMP-CPP) are not active in supporting translocation (CHEN and TAI 1985). While dATP supports protein translocation as efficiently as ATP, other nucleotides, including deoxynucleotides and ADP, can partially replace ATP for translocation. However, the translocation requires either ATP or dATP specifically: the activity by other nucleotides is apparently due to their conversion to ATP as shown by inhibition of translocation upon depletion of ATP by *E. coli* glycerol kinase or in the absence of exogeneous Mg^{2+} (CHEN and TAI 1985; in preparation). In the absence of Mg^{2+} protein translocation can still occur with ATP or dATP. Although gamma-thio-ATP is not active, alpha-thio-ATP is active, albeit slowly, in supporting protein translocation. Azido-ATP, etheno-

ATP, adenosine-tetraphosphate, and phosphoenopyruvate are all not active (Chen and Tai, manuscript in preparation). These results suggest that translocation exhibits specificity both for the adenosine and the phosphate moiety of ATP.

A requirement of ATP for post-translational translocation of protein has also been reported for chloroplasts (GROSSMAN et al. 1980). The translocation of proteins into mitochondria, in the opposite direction to that in the bacterial membrane, has been reported to require protonmotive force (GASSER et al. 1982; SCHLEYER et al. 1982); whether the protonmotive provides energy directly or acts by generating ATP is not known. The endoplasmic reticulum of eukaryotic cells cannot generate a protonmotive force, and even though theoretical calculations have suggested that the energy of peptide chain elongation or peptide folding could be adequate to support translocation (VON HEIJNE and BLOMBERG 1979; ENGELMEN and STEITZ 1981), it may well be that ATP is required. The source of energy for protein translocation in eukaryotic and prokaryotic cells may thus be the same.

2.5 An ATP-Binding Site in the Membrane

Recent work has shown that ATP-driven protein translocation is inhibited competitively by an excess amount of AMP, ADP, nonhydrolyzable ATP analogs, or pyrophosphate and, to a lesser degree, by other nucleotides and deoxynucleotides but not by adenosine, adenosine-tetraphosphate, or NAD^+ (Chen and Tai, in preparation). These results suggest the involvement of at least an ATP-binding site.

That an ATP-binding site in the membrane is involved in protein translocation is supported by the affinity labeling studies with an ATP analog. The photo-crosslinking of azido-ATP or azido-AMP renders the membranes inactive for subsequent translocation in the presence of ATP, even after the unreacted analogs have been removed. Inactivation is prevented by the presence of ATP or AMP during photoreaction, indicating a specific effect on an ATP-binding site in the membrane.

The existence of an ATP-binding site suggests the involvement of membrane protein(s) in protein translocation. Whether these proteins correspond to those that are degraded by treatment of membranes with protease (RHOADS et al. 1984; CHEN et al. 1985) or those identified genetically (Beckwith and Ferro-Novick, this volume) remains to be determined.

2.6 Lipid-Protein Interaction

Several membrane perturbants, including ethanol, phenylethyl alcohol, and procaine, and growth at lower temperatures have been shown to affect secretion of periplasmic proteins and assembly of outer membrane proteins in *E. coli* cells, suggesting the involvement of lipid fluidity in protein export (TRIBHUMAN et al. 1970; TRIBHUMAN and PRADHAN 1977; KIMURA and IZUI 1976; DIRIENZO and INOUYE 1979; PALVA et al. 1981; PAGES et al. 1978; LAZDUNSKI et al. 1979; PAGES and LAZDUNSKI 1982). Similar results have also been observed in the in vitro translocation system (RHOADS et al. 1984; CHEN et al. 1985; see Sect. 2.1). Moreover, several membrane-active agents inhibit the post-translational translocation of alkaline phosphatase and OmpA protein at concentrations that do not disrupt membrane integrity. These include peptide antibiotics polymyxin B, gramicidin S, tyrocidines, and duramycin (Chen and Tai, manuscript in preparation), the latter having been shown to interact with ATPase of clathrin-coated vesicles (STONE et al. 1984; NAKAMURA and RACKER 1984). Taken together these results indicate the importance of the lipid fluidity and the interaction of proteins and lipids in protein translocation. An active role for phospholipids in protein translocation has been suggested earlier (NESMEYANOVA 1982).

Curiously, several higher alcohols, such as tetra- and hexa-decanol and cerulenin, at concentrations which severely impair the synthesis and secretion of extracellular enzymes in *Bacillus* (FISHMAN et al. 1978, 1980), have no inhibitory effect on the in vitro *E. coli* protein translocation. Whether this differential action reflects differences in the role of lipid or fundamental differences in the mechanisms of protein secretion in *E. coli* and *Bacillus* is not clear.

2.7 Limitations and Perspectives

So far, the criterion for the successful in vitro translocation of a protein has been resistance of the products to protease digestion and sedimentation with inverted membrane vesicles, presumably because membranes provide a physical barrier. Although this resistance is lost in the presence of a detergent which solubilizes membranes and translocation is both temperature and ATP dependent (CHEN et al. 1985; CHEN and TAI 1985), the criterion of resistance to protease should be viewed with caution, especially since many secreted proteins are intrinsically resistant to proteases. Furthermore, it has not yet been possible to determine whether the protease resistance is due to complete translocation across or merely insertion into the cytoplasmic membrane. Moreover, the final localization of an outer membrane protein such as OmpA protein clearly requires additional steps. Finally, protein translocation into membrane vesicles, like all other in vitro biological systems, must take cognizance of its much slower kinetics, lower efficiency, and the fact that it represents only a partial reaction of the fast and complex chain of events that occurs in vivo.

Despite these limitations, the in vitro system provides a promising approach for the biochemical analysis of the components, both in the cytoplasm and in the membrane, that are involved in protein translocation. The dissociation of translocation from translation further simplifies the analysis: it is now clear that ATP is essential for protein translocation, and that at least an ATP-binding site and lipid moiety in the membrane are involved (Fig. 1). Hopefully, these findings will provide a basis for further characterization of the components involved. In addition, by using purified, competent precursor, it should be possible to identify cytoplasmic factor(s) involved in translocation. Preliminary experiments with a partially purified precursor of OmpA protein have revealed the requirement for a factor(s) in the soluble cytoplasmic fraction; whether it is identical with the export factor of MÜLLER and BLOBEL (1984b; this volume) is not clear. Moreover, whether the existence of several signal peptidases (Ray et al., Wu and Tokunaga, this volume) and of a mutation that only affects a subset of exported protein (KUMAMOTO and BECKWITH 1985; Beckwith and Ferro-Novcik, this volume) indicate different export pathways for various proteins or only reflect variations of a general pathway, and whether a protein that can be translocated co- or post-translationally (see Sect. 2.2) utilizes the same or alternative export pathway, remains to be elucidated. Another challenge is the elucidation of the role of overall topology and conformation of membrane components, including the effect of protonmotive force, in protein translocation. It is hoped and expected that convergence of biochemical analyses and genetic studies of the components involved in protein translocation will soon clarify the mechanisms of these processes.

3 A Secretory Protein Complex in *B. subtilis*

Many enzymes are secreted by *Bacillus* cells directly into the medium, facilitating their recovery and purification. *Bacilli* have long been important industrially and will be even more so with the prospect of making useful proteins by recombinant DNA technology (see Sarvas, this volume). The single membrane of *Bacilli* provides a simpler model system for the study of the interaction of ribosomes and membranes. The cotranslational secretion of proteins in *Bacilli* has been demonstrated directly by extracellular labeling (SMITH et al. 1979, 1981) and by protease digestion (SMITH et al. 1978) of nascent peptide chains. The experiments on protection from protease digestion have further shown that the protected length is quite uniform for many nascent peptide chains traversing the membrane. The membrane protects a length of 21 amino acid residues for 50% of the chains (SMITH et al. 1978), while the rest of the protected chains are longer and of varying length; they are presumably folded into the membrane or folded as a domain inside the membrane before translocation (RANDALL 1983). The ribosomes thus "shaved" of their extruding chains are still firmly bound to the membranes, suggesting the interaction of peptide-secreting ribosomes with surrounding membrane proteins. KREIBICH et al. (1978a, b) had observed that proteins of the rough and smooth microsomes differ in only two major components that form the ribosome-binding site. As an approach to analyze the membrane proteins that may constitute the "machinery" for protein secretion, we compared in some detail the proteins of ribosome-free and ribosome-bearing membranes.

3.1 Identification of a 64-kD Protein Covered by Secreting Ribosomes

To identify specific membrane proteins involved in protein secretion, the cytoplasmic membrane fragments of *B. subtilis* were separated by sucrose gradient centrifugation into a ribosome-bearing and a ribosome-free fraction. Analysis of these membrane fractions by gel electrophoresis revealed that each fraction contains several unique proteins (Marty-Mazars et al. 1983; Caulfield et al. 1983). Antibodies to several proteins that are unique to ribosome-bearing membranes were prepared and used to study their topography in the membrane (Horiuchi et al. 1983a, b). Of particular interest is the 64-kD protein, which is located on the cytoplasmic surface of the membrane and is not accessible to protease digestion or to its antibody until ribosomes have been released. The tight coverage of this protein by ribosome suggests that it is part of an apparatus of secretion, located on the cytoplasmic surface. In addition, the 64-kD protein is also present in the cytoplasm, suggesting a cyclical attachment to the membrane.

3.2 The 64-kD Protein is Also Present As a Multiprotein Complex

The 64-kD protein has in turn provided a handle for identifying additional proteins that may also be involved in secretion. Immunoprecipitation of a subcellular fraction with antibody to this protein, either in the presence of nondenaturating detergent or in the presence of sodium dodecyl sulfate (SDS) after prior chemical cross-linking, yielded a complex of four proteins (64, 60, 41, and 36 kD). This "S-complex" was found to be associated with membrane-free 70S ribosome, rather than with the membrane (Caulfield et al. 1984). In addition, the 64-kD protein was also present, without the other proteins of the S-complex, in the membrane-ribosome fraction and in the cytosol. Only the form present in the membrane-ribosome fraction was protected from protease digestion. These findings suggest a cycle in which the S-complex participates in the early step of secretion but not in the later stages. It is not yet clear whether the 64-kD protein found in the membrane-ribosome complexes is retained on the membrane after the rest of the complex is released and later recycles via the cytosol, or whether it is a separate pool. Curiously, even though the 64-kD protein is associated (in the complex) with membrane-free ribosomes, in the membrane-ribosome it is firmly associated (without the S-complex) in the membrane (Horiuchi et al. 1983a).

Adler and Arvidson (1984) have recently reported similar findings in another gram-positive bacterium, *Staphylococcus aureus*: antibody to a 60-kD protein (presumably from membrane) of that organism coprecipitates a set of proteins. Moreover, it also precipitates the 64-kD protein and the associated complex of *B. subtilis*. A protein complex associated with SecA protein in *E. coli* has also been noted (Oliver and Beckwith 1981). Furthermore, antibody to the 64-kD protein of *B. subtilis* precipitates a complex of proteins from an *E. coli* extract, some of which correspond in size to those precipitated by SecA antibody (unpublished observation). It is not clear whether these complexes are related structurally or functionally to the signal recognition particle of eukaryotic cells (Walter and Blobel 1981).

3.3 The S-Complex Forms an Organized Structure When Released from the Ribosome

The S-complex of *B. subtilis* can be shown to be attached to the membrane-free 70S ribosomes: it can be chemically cross-linked to ribosomal proteins, and antibody to the 64-kD protein can bind and aggregate these ribosomes, as observed by electron microscopy (CAULFIELD et al. 1985). Low Mg^{2+} or prolonged high-speed centrifugation in a sucrose gradient releases the S-complex from the ribosomes as an aggregate with an S value of 76. Moreover, after release but not in the monosome fraction, the S-complex is seen in the electron microscope as a specific, organized structure, larger than the ribosome. It is intriguing that this structure somewhat resembles the clathrin cages of eukaryotic cells, which have been implicated in membrane protein transport (SALISBURY et al. 1983). Although these organized structures may be artifactual aggregates, their homogeneity suggests that they may have a physiological function.

3.4 Perspectives

The observation of a cage-like structure, formed from a multiprotein complex that is evidently involved in protein secretion in *B. subtilis*, raises the question of whether such a structure is functional in bacteria, in a manner similar to the involvement of clathrin in protein secretory pathways in eukaryotic cells. Further work is needed to establish the existence of this structure in cells and to define its function in an in vitro translocation system. So far, we have not been able to detect any effect of this *B. subtilis* S-complex on the heterologous *E. coli* translocation system (Sect. 2). It may well be that the secretion machinery in *E. coli* and in *B. subtilis* is sufficiently different (see Sarvas, this volume). A homologous *B. subtilis* translation-translocation system is being developed to define the functions of the 64-kD protein and the S-complex. Moreover, while genetic studies on protein localization in *E. coli* have been extensive, relatively little is known in *B. subtilis*. It would be desirable to have genetic information on the components thus far identified biochemically and presumably parts of secretion machinery for the better understanding of mechanisms of protein secretion.

4 Concluding Remarks

Biochemical analyses, both with an in vitro *E. coli* translocation system and with interaction of ribosomes and membranes of *B. subtilis*, have revealed several components that are involved in protein translocation. Each has opened up new approaches for further experimentations to unravel the molecular mechanisms of protein secretion in these bacteria. However, it is not yet possible to correlate the findings of each system, and it is not clear whether two systems are different. It may well be that bacteria possess diverse mechanisms for protein export while others, e.g., mitochondria, chloroplasts, and dog pancreatic membranes, have evolved into specialized systems, despite some overall similarity. The prospects of combining biochemical and genetics studies, both in vivo and in vitro, give rise to the hope that at least some aspects of the bacterial secretion mechanism will soon become clear.

Gram-negative bacteria provide an additional challenge: how are proteins excreted past a double membrane into the extracellular medium during growth? While *E. coli* excretes few proteins (except in the case of specialized systems, see Mackman et al., this volume; de Graaf and Oudega, this volume), others such as *Pseudomonas aeruginosa*, *Aeromonas hydrophila*, and *Vibrio cholerae* excrete many proteins into the medium without an apparent periplasmic pool. We have proposed a possible mechanism in which the protein is laterally transferred from the inner surface of the cytoplasmic membrane through the Bayer junction to the exterior of the outer membrane, from which it is released into the medium (LORY et al. 1983). This mechanism needs to be further substantiated and may also require additional gene products. Alternatively, proteins may be excreted through a periplasmic pool either as precursors or mature size molecules (FECYCZ and CAMPBELL 1985; HIRST et al. 1984; HOWARD and BUCKLEY 1985a, b), the detailed mechanisms of which also need to be elucidated.

Acknowledgments. I thank all my colleagues who have contributed toward the work described here, Henry Paulus for reading the manuscript and making suggestions, and Dotty Syrigos for typing it. This work has been supported by grants from the National Institutes of Health.

Note added after completion of the manuscript:

The lack of translational arrest (see Sect. 2.2.1) by signal recognition particle has recently been reported with Hela cell extracts or with reticulocyte extracts to synthesize secreting protein (MEYER 1985). If this could be further substantiated, translational arrest may not be an important step in bacterial and in mammalian protein secretion, at least in vitro.

References

Adler LA, Arvidson S (1984) Immunological crossreaction between proteins supposed to be involved in protein secretion in *Staphylococcus aureus*. FEMS Microbiol Lett 23:17–20

Bakker EP, Randall LL (1984) The requirement for energy during export of β-lactamase in *Escherichia coli* is fulfilled by the total protonmotive force. EMBO J 3:895–900

Benson S, Hall MN, Silhavy TJ (1985) Genetic analysis of protein export in *Escherichia coli* K12. Annu Rev Biochem 54:101–134

Caulfield MP, Tai PC, Davis BD (1983) Association of penicillin binding proteins and other enzymes with the ribosome-free membrane fraction of *Bacillus subtilis*. J Bacteriol 156:1–5

Caulfield MP, Horiuchi S, Tai PC, Davis BD (1984) The 64-kilodalton membrane protein of *Bacillus subtilis* is also present as a multiprotein complex on membrane-free ribosomes. Proc Natl Acad Sci USA 81:7772–7776

Caulfield MP, Furlong D, Tai PC, Davis BD (1985) The secretory S complex of *Bacillus subtilis* forms a large, organized structure when released from ribosomes. Proc Natl Acad Sci USA 82:4031–4035
Chang CN, Model P, Blobel G (1979) Membrane biogenesis: cotranslational integration of the bacteriophage f_1 coat protein into an *Escherichia coli* membrane fraction. Proc Natl Acad Sci USA 76:1251–1255
Chen LL, Tai PC (1985) ATP is essential for protein translocation into *Escherichia coli* membrane vesicles. Proc Natl Acad Sci USA 82:4384–4388
Chen LL, Rhoads D, Tai PC (1985) Alkaline phosphatase and OmpA protein can be translocated posttranslationally into membrane vesicles of *Escherichia coli*. J Bacteriol 161:973–980
Daniels CJ, Anderson JJ, Landick R, Oxender DL (1980) The in vitro synthesis and processing of the branched-chain amino acid binding proteins. J Supramol Struct 14:305–311
Daniels CJ, Bole DG, Quay SC, Oxender DL (1981) A role for membrane potential in the secretion of protein into the periplasm of *Escherichia coli*. Proc Natl Acad Sci USA 77:4669–4673
Date T, Wickner W (1981) Isolation of the *Escherichia coli* leader peptidase gene and effects of leader peptidase overproduction *in vivo*. Proc Natl Acad Sci USA 78:6106–6110
Davis BD, Tai PC (1980) The mechanism of protein secretion across membranes. Nature 283:433–437
DiRienzo JM, Inouye M (1979) Lipid fluidity-dependent biosynthesis and assembly of the outer membrane proteins of *E. coli*. Cell 17:155–161
Enequist HG, Hirst TR, Hardy SJS, Harayama S, Randall LL (1981) Energy is required for maturation of exported proteins in *Escherichia coli*. Eur J Biochem 116:227–233
Engelmen DM, Steitz TA (1981) The spontaneous insertion of proteins into and across membranes: the helical hairpin hypothesis. Cell 23:411–422
Fecycz IT, Campbell N (1985) Mechanisms of activation and secretion of a cell-associated precursor of an exocellular proteases of *Pseudomonas aeruginosa* 34362A. Eur J Biochem 146:35–42
Fishman Y, Rottem S, Citri N (1978) Evidence linking penicillinase formation and secretion to lipid metabolism in *Bacillus licheniformis*. J Bacteriol 134:434–439
Fishman Y, Rottem S, Citri N (1980) Preferential suppression of normal exoenzyme formation by membrane-modifying agents. J Bacteriol 141:1435–1438
Garwin JL, Beckwith J (1982) Secretion and processing of ribose-binding protein in *Escherichia coli*. J Bacteriol 149:789–792
Gasser SM, Daum G, Schatz G (1982) Import of proteins into mitochondria. J Biol Chem 257:13034–13041
Gilmore R, Walter P, Blobel G (1982) Protein translocation across the endoplasmic reticulum. II. Isolation and characterization of the signal recognition particle receptor. J Cell Biol 95:470–477
Goodman JM, Watts C, Wickner W (1981) Membrane assembly: posttranslational insertion of M13 procoat protein into *E. coli* membranes and its proteolytic conversion to coat protein *in vitro*. Cell 24:437–444
Grossman A, Bartlett S, Chua NH (1980) Energy-dependent uptake of cytoplasmically synthesized polypeptides by chloroplasts. Nature 285:625–628
Hirst TR, Sanchez J, Kaper JB, Hardy SJ, Holmgren J (1984) Mechanism of toxin secretion by *Vibrio cholerae* investigated in strains harboring plasmids that encode heat-labile enterotoxins of *Escherichia coli*. Proc Natl Acad Sci USA 81:7752–7756
Horiuchi S, Marty-Mazars D, Tai PC, Davis BD (1983a) Localization and quantitation of proteins characteristic of the complexed membrane of *Bacillus subtilis*. J Bacteriol 154:1215–1221
Horiuchi S, Tai PC, Davis BD (1983b) A 64-kilodalton membrane protein of *Bacillus subtilis* covered by secreting ribosomes. Proc Natl Acad Sci USA 80:3287–3291
Howard SP, Buckley JT (1985a) Protein export by a gram-negative bacterium: production of aerolysin by *Aeromonas hydrophila*. J Bacteriol 161:1118–1124
Howard SP, Buckley JT (1985b) Activation of the hole-forming toxin aerolysin by extracellular processing. J Bacteriol 163:336–340
Innis MA, Tokunaga M, Williams ME, Loranger JM, Chang SY, Chang S, Wu HC (1984) Nucleotide sequence of *Escherichia coli* prolipoprotein signal peptidase (*lsp*) gene. Proc Natl Acad Sci USA 81:3708–3712
Josefsson L, Randall LL (1981) Different exported proteins in *Escherichia coli* show differences in the temporal mode of processing *in vivo*. Cell 25:151–157
Kimura K, Izui K (1976) Importance of membrane fluidity in the induction of alkaline phosphatase, a periplasmic enzyme in *Escherichia coli*. Biochem Biophys Res Commun 70:900–906

Koshland D, Botstein D (1982) Evidence for post-translational translocation of beta-lactamase across the bacterial inner membrane. Cell 30:893–902

Kreibich G, Czako-Graham M, Grebenau R, Mok W, Rodriguez-Boulan E, Sabatini D (1978a) Characterization of the ribosomal binding site in rat liver rough microsomes: ribophorins I and II, two integral membrane proteins related to ribosome binding. J Supramol Struct 8:279–302

Kreibich G, Ulrich BL, Sabatini D (1978b) Proteins of rough microsomal membranes related to ribosome binding I. Identification of ribophorins I and II, membrane proteins characteristic of rough microsomes. J Cell Biol 77:464–487

Kumamoto C, Beckwith J (1985) Evidence for specificity at an early step in protein export in *Escherichia coli*. J Bacteriol 163:267–274

Lazdunski C, Baty D, Pages JM (1979) Procaine, a local anesthetic interacting with the cell membrane, inhibits the processing of precursor forms of periplasmic proteins in *Escherichia coli*. Eur J Biochem 96:49–57

Lin JJ-C, Kanazawa H, Ozols J, Wu HC (1978) *Escherichia coli* mutant with an amino-acid alteration with signal sequence of outer membrane prolipoprotein. Proc Natl Acad Sci USA 75:4891–4895

Lory S, Tai PC, Davis BD (1983) Mechanisms of protein excretion by gram-negative bacteria: *Pseudomonas aeruginosa* exotoxin A. J Bacteriol 156:695–702

Marty-Mazars D, Horiuchi S, Tai PC, Davis BD (1983) Proteins of ribosome-bearing and -free membrane domains in *Bacillus subtilis*. J Bacteriol 154:1381–1388

Meyer DI (1985) Signal recognition particle (SRP) does not mediate a translational arrest of nascent secretory proteins in mammalian cell-free systems. EMBO J 4:2031–2033

Meyer DI, Krause E, Dobberstein B (1982) Secretory protein translocating across membranes – the role of the "docking protein." Nature 297:647–650

Michaelis S, Beckwith J (1982) Mechanism of incorporation of cell envelope proteins in *Escherichia coli*. Annu Rev Microbiol 36:435–465

Müller M, Blobel G (1984a) In vitro translocation of bacterial proteins across the plasma membrane of *Escherichia coli*. Proc Natl Acad Sci USA 81:7421–7425

Müller M, Blobel G (1984b) Protein export in *Escherichia coli* requires a soluble activity. Proc Natl Acad Sci USA 81:7737–7741

Nakamura S, Racker E (1984) Inhibitory effect of duramycin on partial reactions catalyzed by (Na^+, K^+)-adenosine triphosphatase from dog kidney. Biochemistry 23:385–389

Nesmeyanova MA (1982) On the possible participation of acid phospholipids in the translocation of secreted proteins through the bacterial cytoplasmic membrane. FEBS Lett 142:189–193

Oliver D, Beckwith J (1981) *E. coli* mutant pleiotropically defective in the export of secreted proteins. Cell 25:765–772

Osborn MJ, Wu H (1980) Proteins of the outer membrane of Gram-negative bacteria. Annu Rev Microbiol 34:369–422

Pages JM, Lazdunski C (1982) Maturation of exported proteins in *Escherichia coli*. Requirement of energy, site and kinetics of processing. Eur J Biochem 124:561–566

Pages JM, Piovant M, Varronne S, Lazdunski C (1978) Mechanistic aspects of the transfer of nascent periplasmic proteins across the cytoplasmic membranes in *Escherichia coli*. Eur J Biochem 80:589–602

Pages JM, Anba J, Bernadac A, Shinagawa H, Nakata A, Lazdunski C (1984) Normal precursors of periplasmic proteins accumulated in the cytoplasm are not exported posttranslationally in *Escherichia coli*. J Biochem 143:499–505

Palva ET, Hirst TR, Hardy SJS, Holmgren J, Randall LL (1981) Synthesis of a precursor to the B subunit of heat-labile enterotoxin in *Escherichia coli*. J Bacteriol 146:325–340

Randall LL (1983) Translocation of domains of nascent periplasmic proteins across the cytoplasmic membrane is independent of elongation. Cell 33:231–240

Randall LL, Hardy SJS (1984) Export of protein in bacteria. Microbiol Rev 48:290–298

Rhoads DB, Tai PC, Davis BD (1984) Energy-requiring translocation of the OmpA protein and alkaline phosphatase of *Escherichia coli* into membrane vesicles. J Bacteriol 159:63–70

Salisbury JL, Condeelis JS, Satir P (1983) Receptor-mediated endocytosis: machinery and regulation of the clathrin-coated vesicle pathway. Int Rev Exp Pathol 24:2–63

Schleyer M, Schmidt B, Neupert W (1982) Requirement of a membrane potential for the posttranslational transfer of proteins into mitochondria. Eur J Biochem 125:109–116

Silhavy TJ, Benson SA, Emr SD (1983) Mechanisms of protein localization. Microbiol Rev 47:313–344

Smith WP (1980) Cotranslational secretion of diphtheria toxin and alkaline phosphatase *in vitro*: involvement of membrane protein(s). J Bacteriol 141:1142–1147
Smith WP, Tai PC, Thompson RC, Davis BD (1977) Extracellular labeling of nascent polypeptides traversing the membrane of *Escherichia coli*. Proc Natl Acad Sci USA 74:2830–2834
Smith WP, Tai PC, Davis BD (1978) Interaction of secreted chains with surrounding membrane in *Bacillus subtilis*. Proc Natl Acad Sci USA 75:814–817
Smith WP, Tai PC, Davis BD (1979) Extracellular labeling of growing secreted polypeptide chains in *Bacillus subtilis* with diazoiodosulfanic acid. Biochemistry 18:198–203
Smith WP, Tai PC, Davis BD (1981) *Bacillus licheniformis* penicillinase: cleavages and attachment of lipid during cotranslational secretion. Proc Natl Acad Sci USA 78:3501–3505
Stone DK, Xie X-S, Racker E (1984) Inhibition of clathrin-coated vesicle acidification by duramycin. J Biol Chem 259:2701–2703
Tribhuwan RC, Pradhan DS (1977) Induction of alkaline phosphatase in *Escherichia coli*: effect of procaine-hydrochloride. J Bacteriol 131:431–437
Tribhuwan RC, Pilgaokar AK, Pradhan DS, Screenivasan DS (1970) Effect of phenethylalcohol on induction of alkaline phosphatase in *Escherichia coli*. Biochem Biophys Res Commun 41:244–250
Varenne S, Piovant M, Pages JM, Lazdunski C (1978) Evidence for synthesis of alkaline phosphatase on membrane-bound polysomes in *Escherichia coli*. Eur J Biochem 86:603–606
Von Heijne G, Blomberg C (1979) Transmembrane translocation of protein. Eur J Biochem 97:175–181
Walter P, Blobel G (1980) Purification of a membrane-associated protein complex required for protein translocation across the endoplasmic reticulum. Proc Natl Acad Sci USA 77:7112–7116
Walter P, Blobel G (1981) Translocation of proteins across the endoplasmic reticulum. III. Signal recognition protein (SRP) causes signal-sequence dependent and site-specific arrest of chain elongation that is released by microsomal membranes. J Cell Biol 91:557–561
Wickner W (1980) Assembly of proteins into membranes. Science 210:861–868
Yu F, Yamada H, Daishima K, Mizushima S (1984) Nucleotide sequence of the *lspA* gene, the structural gene for lipoprotein peptidase of *Escherichia coli*. FEBS Lett 173:264–268
Zimmermann R, Wickner W (1983) Energy and intermediates of the assembly of protein OmpA into the outer membrane of *Escherichia coli*. J Biol Chem 258:3920–3925
Zwizinski C, Wickner W (1980) Purification and characterization of leader (signal) peptidase from *Escherichia coli*. J Biol Chem 255:7973–7977

Secretion Cloning Vectors for Guiding the Localization of Proteins in Vivo

C.A. Lunn[1], M. Takahara[2], and M. Inouye[1]

1 Introduction

Cloning of genes into high copy number vectors has supplied biochemists with a powerful method for the production and purification of large amounts of biochemically important macromolecules. Judicious choice of promoter sequences has produced vectors capable of loading cells with 50% of the total cellular protein as the desired gene product, making purification a trivial chore. However, this approach is limited to proteins whose activity is not detrimental to protein production, and to proteins not sensitive to cytoplasmic proteases (see below). Recognition of this problem has stimulated the development of several cloning vectors designed to guide a cloned gene product out of the cytoplasm of the bacterial cell. These vectors produce a hybrid protein in which the cloned gene product is fused with a signal peptide, a structural element necessary for protein translocation. In this review, we will describe the advantages and limitations of some currently available secretion cloning vectors, and will discuss the use of these tools to produce large amounts of protein and

[1] Department of Biochemistry, State University of New York, Stony Brook, NY 11794, USA
[2] Biochemical Research Laboratory, Tokyo Jozo Co. Ltd., 632-1 Mifuku, Ohito-Cho, Tagata-Gun, Shizuoka-Ken, Japan 410-23

Current Topics in Microbiology and Immunology, Vol. 125

to determine the information required to direct proteins to the various compartments of the cell.

2 Importance of Secretion Cloning Vectors

The value of secretion cloning vectors derives from the importance of positioning certain cloned gene products in one of the different compartments of the bacterial cell. The cytoplasm, the site of synthesis of a cloned gene product, also contains the major biochemical reactions involved in cell metabolism. Thus, expression of cloned gene products within this compartment is limited to those that do not interrupt cell metabolism or kill the host cell. Secretion cloning vectors permit cloning of nucleases and proteases which would otherwise be difficult. Second, many cloned gene products (for example, small polypeptide hormones) are sensitive to the high proteolytic activity present in the bacterial cytoplasm. Secretion cloning vectors allow transport of these hormones to regions protected from cytoplasmic proteases, such as the periplasmic space, membrane, or growth medium. Third, the cytoplasm contains the majority of the protein present in the bacterial cell. Therefore, one can achieve severalfold purification of the cloned gene product via its secretion into the periplasmic space or the culture medium, allowing production of highly purified soluble products. Finally, production of proteins in the cytoplasm requires an initiator methionine at the amino terminus. However, most of the proteins that one wishes to produce in bacteria do not have the amino terminal methionine. Secretion vectors allow the production of a product with a desired amino terminal amino acid residue. For all of these reasons, secretion cloning vectors provide powerful tools for the production of biologically important proteins.

3 Characteristics of Secretion Cloning Vectors

3.1 Characteristics of Expression Vehicle

Plasmids forming the basis of the currently available secretion cloning vectors possess several common characteristics. They are generally high copy number plasmids carrying one or more highly efficient and stringently controlled promoters. These two factors insure production of large amounts of protein from the cloned gene while the design allows expression of the cloned gene product to be controlled. Thus, the cloned gene product may be produced only under desired growth conditions (e.g., when amplification of the cloned gene products is least detrimental to the host cell metabolism).

3.2 Structural Requirement for Secretory Proteins

Secretion cloning vectors are designed to produce a hybrid cloned gene product containing a peptide extension of the amino terminus of the cloned gene product.

The added sequence includes the signal peptide, a peptide present on the biosynthetic precursors of secretory proteins in vivo (SILHAVY et al. 1983). This peptide is involved in the initial interaction between the membrane and the translocated protein (INOUYE and HALEGOUA 1980) and requires certain structural characteristics to function efficiently as a signal peptide (DUFFAUD et al. 1985). In bacteria, the amino terminus of the signal peptide must contain one to three positively charged amino acids. This is followed by a hydrophobic domain consisting of 14–20 amino acid residues, with centrally located proline or glycine residues. Serine and/or threonine are usually located on the carboxyl terminus of this hydrophobic domain. Finally, correct processing by the signal peptidase to cleave off this extension and release the protein to the outside of the cytoplasmic membrane requires that the carboxy terminal residue of the signal peptide be either alanine or glycine. Together, the signal sequence is proposed to form a hydrophobic loop that partitions into the lipid bilayer to lead the translocation of the protein across the membrane (INOUYE and HALEGOUA 1980).

4 Characteristics of the Cloned Gene Required for Translocation: Current Limitations to Secretion Cloning

The presence of a signal peptide is necessary to direct a protein into the cytoplasmic membrane, and to translocate the amino terminus of the protein from the interior to the exterior surface of the cytoplasmic membrane (INOUYE and HALEGOUA 1980). The ultimate localization of the protein within the cytoplasmic membrane, periplasm, or outer membrane requires structural characteristics present within the protein (INOUYE and HALEGOUA 1980). Thus, cytoplasmic β-galactosidase cannot be converted into an outer membrane protein simply by fusing this hydrophilic protein with the signal sequence of *lamB*, an outer membrane protein (MORENO et al. 1980). Similarly, the signal sequence from a periplasmic protein cannot direct a hydrophobic protein into the periplasm (TOMMASSEN et al. 1983). This suggests that directing a protein to a desired compartment may require fusing the protein with critical portions of a representative protein from that compartment. The secretion cloning vectors are efficient at directing normally secreted proteins to either the periplasm or the outer membrane. These vectors are also efficient at translocating small molecular weight peptides to the periplasm or outer membrane (for examples, see below). However, although these vectors may convert any normally cytoplasmic protein into a membrane-associated protein (associated with the inner surface of the cytoplasmic membrane), these vector systems have a limited ability to translocate cytoplasmic proteins across the cytoplasmic membrane. Translocation may be halted by a stop-transfer sequence within the polypeptide chain or by a stable conformation formed by the polypeptide within the cytoplasm before its translocation. An understanding of this structural information will assist in predicting the utility of secretion vectors in transport.

4.1 Stop-Transfer Sequence

One strategy used to interrupt translocation of an integral membrane protein across a membrane is via a stop-transfer sequence (Blobel 1980). This peptide contains similar sequence characteristics to the signal peptide, consisting of a stretch of hydrophobic amino acids flanked by positively charged residues. The sequence differs from the signal sequence in its location within the protein, with the stop-transfer sequence distant from the amino terminus of the protein. When this sequence encounters the membrane bilayer, it forms a stable interaction which arrests further translocation of the polypeptide. Peptides with these characteristics have been observed within the sequences of several integral membrane proteins (Rose et al. 1980; Yost et al. 1983).

The stop-transfer sequence is believed to act independently to the translocation machinery, forming a stable structure with the lipid bilayer itself. Also, the stop-transfer sequence does not act in the absence of a signal peptide. With no recognition process involved in the activity of this peptide, any sequence consisting of a hydrophobic stretch of sufficient length flanked by charged residues should act as a stop-transfer sequence. Because most cytosolic proteins contain hydrophobic sequences buried within them (Schultz and Schirmer 1979), such sequences may act as stop-transfer sequences when cloned into secretion cloning vectors, resulting in the formation of membrane-bound protein. Computer modeling (von Heijne 1980) has supported this prediction, concluding that hydrophobic stretches present in cytoplasmic β-galactosidase inhibit translocation of a β-lactamase- β-galactosidase hybrid protein. This result is experimentally observed in vivo (Silhavy et al. 1979; Fowler and Zabin 1978).

4.2 Protein Folding

A second characteristic that could block translocation of a cloned protein through a membrane bilayer involves the stability of the folded protein. The translocation of some bacterial proteins through the membrane occurs via a post-translational mechanism (Randall and Hardy 1984). By this mechanism, the entire polypeptide chain is synthesized before translocation through the membrane. This contrasts with cotranslational translocation, by which the polypeptide threads through the bilayer as it comes off the ribosome. The post-translational translocation process presumably involves local uncoiling of each domain of the folded protein, followed by linear transfer of the polypeptide chain through the membrane bilayer. Secreted proteins are sufficiently unstable to allow this reaction to occur. It is likely that cloned proteins also remain in the cytoplasm for sufficient time to allow folding at the inner surface of the cytoplasmic membrane. The bacterial translocation system therefore must provide adequate energy to destabilize the folded domains of the cloned protein before translocation. Failure to accomplish this would result in the cloned protein precursor being trapped at the inner surface of the cytoplasmic membrane.

5 Examples of Secretion Cloning Vectors

5.1 Secretion in *Bacillus subtilis*

Bacilli are important hosts for the cloning and expression of foreign genes. These gram-positive organisms contain only a single membrane. Interest in this system derives from the ability of these bacteria to secrete large amounts of enzymes into the culture medium. The ability to direct a cloned gene product into the growth medium would make recovery and purification of the product a simple task. However, because its own proteases are also secreted into the medium, the protection of a foreign gene product from proteolytic digestion becomes a formidable problem.

Several secretion vectors have been developed for use with *Bacillus subtilis*. A popular system utilizes the α-amylase promoter and signal sequence, either from *B. amyloliquefaciens* (PALVA et al. 1982) or from *B. subtilis* (OHMURA et al. 1984). The advantage of the α-amylase system is the large amount of synthesis of the α-amylase protein in vivo (approaching 1 g/liter, Table 1) and the ability of the protein to be synthesized in the absence of other protein synthesis (GOULD et al. 1975). Thus, large amounts of α-amylase are produced in stationary-phase

Table 1. Production and secretion of selected proteins in *Bacillus subtilis* and *Escherichia coli*

Promoter	Signal sequence	Protein	Synthesis (mg/liter)	Localization	Ref.
Bacillus subtilis:					
Bacillus amyloliquefaciens		*B. amyloliquefaciens*			
α-amylase	α-amylase	α-Amylase	1000	Medium	ULMANEN et al. (1985)
α-amylase	α-amylase	TEM-β-lactamase	10	Medium	ULMANEN et al. (1985)
α-amylase	α-amylase	SF virus E1	0.1	Medium	ULMANEN et al. (1985)
α-amylase	α-amylase	VSV glycoprotein	0.1	Intracellular	LANDSTROM (1984)
α-amylase	α-amylase	Human α_2 interferon	1	Medium	PALVA et al. (1983)
Escherichia coli:					
E. coli	*E. coli*	*Staphylococcus aureus*			
lpp	*omp*A	Nuclease	300[a]	Periplasm	TAKAHARA et al. (1985)
lpp[P]-5	*omp*A	Nuclease	1500[a]	ND	INOUYE and INOUYE (1985)
		E. coli			
lpp	*omp*A	β-lactamase	300[a]	Periplasm	GHRAYEB et al. (1984)

ND, not determined

[a] Values quoted are approximations based on the percentage of total cellular protein observed upon induction of plasmid-encoded gene. The calculation assumes a culture density of 1×10^{10} cells/ml, and a cellular protein concentration of 0.3 mg protein/10^9 cells

culture of the bacteria when few other proteins are being synthesized. Table 1 lists the synthesis and localization of the number of proteins synthesized using this cloning vehicle. To date, TEM-β-lactamase was the most efficiently produced heterologous protein, reaching approximately 10 mg protein/liter culture (ULMANEN et al. 1985). Upon reaching this peak in synthesis, the amount of protein decreased, the result of extracellular proteases. This occurred in a host strain of *B. subtilis* selected to contain low levels of metallo- and serine proteases (SIBAKOV et al. 1984).

A cloning system utilizing the α-amylase promoter and signal sequence from *B. subtilis* was used to synthesize and secrete biologically active mouse interferon-β (SHIROZA et al. 1985). The total amount of protein synthesized was 4000 U/ml, which decreased with continued culture, again the result of extracellular proteases. A secretion vector system has also been constructed using the *B. licheniformis* penicillinase signal sequence (CHANG et al. 1981). Unlike the α-amylase system, penicillinase is synthesized at maximum rates during logarithmic phase of cell growth, when extracellular proteases are secreted at low rates. The hope is that this system will permit secretion of the cloned gene product into a growth medium containing few extracellular proteases. Initial studies cloning the interferon gene into this vector (CHANG et al. 1983) gave low levels of expression, with 10% of the synthesized interferon localized in the growth medium. This result was attributed to degradation of the hybrid protein by cellular proteases, though the modification of the *penP* signal peptide with glycerolipid may inhibit secretion (ULMANEN et al. 1985).

As has been described, one major difficulty encountered upon utilizing the *B. subtilis* as a vehicle for the cloning of foreign genes is the presence of extracellular proteases secreted by the growing bacteria. These proteases degrade the translocated proteins, resulting in drastically decreased yields of the cloned foreign proteins (ULMANEN et al. 1985) or limitations in the secretion of the products. If this problem can be overcome by the generation of protease-free strains of *B. subtilis* the *Bacillus* species will provide a powerful system for the large-scale production of foreign proteins.

5.2 Secretion Vectors in *Escherichia coli*

The study of protein secretion in bacteria has led to many experiments showing that proteins can be secreted using a heterologous signal sequence. For instance, β-lactamase was translocated across the cytoplasmic membrane using the signal sequence of *lamB* (MORENO et al. 1980), the signal of *ompA* (GHRAYEB et al. 1984), or the signal sequence of lipoprotein (INOUYE et al. 1983). Numerous hybrid proteins formed with β-galactosidase have been generated genetically (for review see Beckwith, this issue). These investigations provide a basis for the development of secretion cloning vectors in *E. coli*. The disadvantage of this gram-negative bacteria is that the translocated proteins are not released into the medium, but are retained within the periplasmic space, or are associated with the cytoplasmic or outer cell membrane. The advantage of using *E. coli* is the fact that the cells do not secrete large amounts of peptidases into the

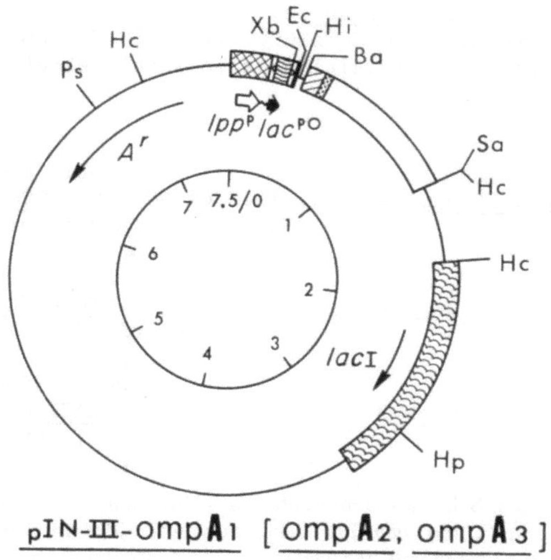

Fig. 1. Schematic representation of secretion vectors pINIII-*ompA*1 (*ompA*2 and *ompA*3). *Ba*, *Bam*HI; *Ec*, *Eco*RI; *Hc*, *Hinc*II; *Hi*, *Hind*III; *Hp*, *Hpa*I; *Ps*, *Pst*I; *Sa*, *Sal*I; *Xb*, *Xba*I, *lpp*^{P}, *lpp* promoter; and *lac*^{PO}, *lac* promoter-operator

growth medium naturally. This alleviates one major problem with the *Bacillus* system. Second, a wealth of information is available on the genetics of *E. coli*. This knowledge allows design of secretion vectors that produce large amounts of foreign protein (see Table 1, and below).

Our interest in the study of protein translocation and the assembly of lipoprotein into the outer membrane of *E. coli* has led to the development of secretion cloning vectors in our laboratory. We have based these vehicles on the pINIII vector system (Masui et al. 1984). This system was constructed from pBR322, a plasmid produced in large amounts (roughly 30 copies) in the bacterial cell (Stueber and Bujard 1982). The pINIII vector utilizes a high efficiency *lpp* promoter, one of the strongest promoters in *E. coli* (Inouye and Halegona 1980). Transcription via this *lpp* promoter is controlled by the *lac* UV5 promoter-operator, inserted downstream from the *lpp* promoter. Therefore, the expression of the cloned gene is repressed by the *lac* repressor (*lacI* gene product) in the absence of a *lac* inducer. The *lacI* gene is inserted into the plasmid, so that expression of the cloned gene is regulated by a *lac* repressor produced by the vector itself. A generalized restriction map of this secretion cloning vector is shown in Fig. 1.

5.2.1 pINIII-ompA Secretion Cloning Vector

We have constructed secretion vectors based on two secreted proteins in *E. coli*. The first utilizes the signal sequence of *ompA* which codes for OmpA, a major outer membrane protein. The *ompA* signal peptide sequence was prepared from a 65-bp *Hha*I fragment from the cloned *ompA* gene (P.J. Green and M. Inouye, unpublished results) plus two synthetic oligonucleotides coding for the Shine-Dalgarno ribosome binding site (Shine and Dalgarno 1975) and the *ompA* signal peptide cleavage site. This vector (pINIII-*ompA*3) has three

```
OmpA signal sequence:
..ATGAAAAAGACAGCTATCGCGATTGCAGTGGCACTGGCTGGTTTCGCTACCGTAGCGCAGGCCGCTCCG...
..MetLysLysThrAlaIleAlaIleAlaValAlaLeuAlaGlyPheAlaThrValAlaGlnAlaAlaPro...
  1                        10                                20  ▲


                        EcoRI  HindIII  BamHI
pINIII-ompA 1   ..GTAGCGCAGGCCGCGAATTCCAAGCTTGGATCCGGCTG...
                ..ValAlaGlnAlaAlaAsnSerLysLeuGlySerGly.....
                             ▲
                         EcoRI  HindIII  BamHI
       ompA 2   ..GTAGCGCAGGCCGCTGAATTCCAAGCTTGGATCCGGCTG...
                ..ValAlaGlnAlaAlaGluPheGlnAlaTrpIleArg......
                             ▲
                        EcoRI  HindIII  BamHI
       ompA 3   ..GTAGCGCAGGCCGGAATTCCAAGCTTGGATCCGGCTG...
                ..ValAlaGlnAlaGlyIleProSerLeuAspPro.......
                          20 ▲
```

Fig. 2. Schematic representation of the linker sequences of secretion cloning vectors pINIII-*ompA*1 (*ompA*2 and *ompA*3). Linker polynucleotides encoding multiple cloning sites are *underlined*. The *arrow* represents the cleavage site of proOmpA by signal peptidase. *Numbers under amino acids* represent position in the OmpA sequence

unique cloning sites, *Eco*RI, *Hin*dIII, and *Bam*HI, immediately after the sequence of the *ompA* signal peptide. In order to express any foreign DNA fragments with any reading frame, two other vectors (pINIII-*ompA*1 and pINIII-*ompA*2) have been constructed. These two vectors are identical to pINIII-*ompA*3, but have the multiple cloning sites in two different reading frames (see Fig. 2). The resulting junctional sequences for each vector, containing the unique *Eco*RI, *Hin*dIII, and *Bam*HI cloning sites, is shown in Fig. 2. Because of these junctional sequences, each cloned gene product contains several extra amino acids at its amino terminal end. However, the sequences coding for these extra amino acids can be easily removed by oligonucleotide-directed site-specific mutagenesis (Morinaga et al. 1984).

5.2.2 pINIII-B and pINIII-C Secretion Cloning Vectors

A second class of secretion vectors was constructed based on the lipoprotein promoter and signal sequence. The pINIII-A vectors are nonsecretion cloning vectors which contain three unique cloning sites (*Eco*RI, *Hin*dIII, and *Bam*HI) after the initiator methionine of the prolipoprotein signal sequence (Fig. 3). This construction results in the cloned gene having no lipoprotein signal sequence. The pINIII-B vectors contain these restriction sites immediately after the lipoprotein signal sequence. Cloning into these sites results in a hybrid protein containing the lipoprotein signal sequence. This plasmid was designed to be equivalent to the pINIII-*ompA* plasmid, thereby directing the cloned protein into the periplasm. This proved not to be the case, as the lipoprotein signal sequence was rarely removed from the cloned gene. This results in a cloned gene product associated with the outer surface of the cytoplasmic membrane via a signal peptide "tail" (see below).

```
Lipoprotein signal sequence:
..TATTAATAATGAAAGCTACTAAACTGGTACTGGGCGCGGTAATCCTGGGTTCTACTCTGCTGGCAGGTTGCTCCAGCAACGCTAAAATCGATCAGCTGTCTTCT..
           MetLysAlaThrLysLeuValLeuGlyAlaValIleLeuGlySerThrLeuLeuAlaGlyCysSerSerAsnAlaLysIleAspGlnLeuSerSer
            1                           10                            20▲

                                                 EcoRI HindIII  BamHI
pINIII-A1  ..ATCTAGAGGGTATTAATAATGAAAGGGAATTCCAAGCTTGGATCCGGCTACTAA...
                                     MetLysGlyAsnSerLysLeuGlySer...

                                                     EcoRI HindIII  BamHI
       A2  ..ATCTAGAGGGTATTAATAATGAAAGGGAAGGAATTCCAAGCTTGGATCCGGCTA...
                                     MetLysGlyLysGluPheGlnAlaTrpIleArg...

                                                   EcoRI HindIII  BamHI
       A3  ..ATCTAGAGGGTATTAATAATGAAAGGGGGAATTCCAAGCTTGGATCCGGCTACT...
                                     MetLysGlyGlyIleProSerLeuAspPro...
                                      1

                                                   EcoRI HindIII  BamHI
pINIII-B1  ..GCGGTAATCCTGGGTTCTACTCTGCTGGGAATTCCAAGCTTGGATCCGGGCAGGTTGCTCCAGC...
           ..AlaValIleLeuGlySerThrLeuLeuGlyAsnSerLysLeuGluSerGly...

                                                       EcoRI HindIII  BamHI
       B2  ..GCGGTAATCCTGCGTTCTACTCTGCTGGGAAGGAATTCCAAGCTTGGATCCGGGCAGGTTGCTCCAGCAACGCT...
           ..AlaValIleLeuGlySerThrLeuLeuGlyArgAsnSerLysLeuGlySerGly...

                                                     EcoRI HindIII  BamHI
       B3  ..GCGGTAATCCTGGGTTCTACTCTGCTGGGGGAATTCCAAGCTTGGATCCGGGCAGGTTGCTCCAGCAACGCT...
           ..AlaValIleLeuGlySerThrLeuLeuGlyGluPheGlnAlaTrpIleArg...
             10

                                                           EcoRI HindIII  BamHI
pINIII-C1  ..CTGGCAGGTTGCTCCAGCAACGCTAAAATCGATCGGAATTCCAAGCTTGGATCCGGAGCTGTCTTCT...
           ..LeuAlaGlyCysSerSerAsnAlaLysIleAspArgAsnSerLysLeuGlySerGly...
                     ▲

                                                               EcoRI HindIII  BamHI
       C2  ..CTGGCAGGTTGCTCCAGCAACGCTAAAATCGATCGGAAGGAATTCCAAGCTTGGATCCGGAGCTGTCTTCT...
           ..LeuAlaGlyCysSerSerAsnAlaLysIleAspArgLysGluPheGlnAlaTrpIleArg...
                     ▲

                                                             EcoRI HindIII  BamHI
       C3  ..CTGGCAGGTTGCTCCAGCAACGCTAAAATCGATCGGGGAATTCCAAGCTTGGATCCGGAGCTGTCTTCT...
           ..LeuAlaGlyCysSerSerAsnAlaLysIleAspArgGlyIleProSerLeuAspPro...
                  20 ▲
```

Fig. 3. Schematic representation of the linker sequences of pINIII-A, pINIII-B, and pINIII-C secretion cloning vectors. Linker polynucleotides encoding multiple cloning sites are *underlined*. The *arrow* represents the cleavage site of prolipoprotein by signal peptidase. *Numbers under amino acids* represent the position in the prolipoprotein sequence

Localization of a protein within the outer membrane requires more structural information than that contained within the signal sequence (DUFFAUD et al. 1985). Our studies of the major lipoprotein suggested that the signal sequence, plus a short peptide (eight to nine amino acids) from the amino terminus of the lipoprotein, are sufficient to translocate β-lactamase to the outer membrane (GHRAYEB and INOUYE 1984). Based on this result, we constructed the pINIII-C cloning vectors, containing the three unique cloning sites after the *lpp* signal sequence, plus eight amino acids from the *lpp* structural gene (MASUI et al. 1983). As with the pINIII-*ompA* vectors described above, the unique *Eco*RI, *Hin*dIII, and *Bam*HI sites for cloning are constructed in three different reading frames (pINIII-C1, pINIII-C2, and pINIII-C3). The sequences of these vectors at the multirestriction sites are shown in Fig. 3.

To compare these secretion cloning vectors, we have investigated the expression and localization of staphylococcal nuclease A, after cloning the gene into each system (TAKAHARA et al. 1985; Takahara et al., manuscript in preparation). This gene was chosen because of its toxicity to cells when expressed in the cytoplasm. We first determined the amount of protein produced by each plasmid in vivo. Exponentially growing cultures of *E. coli* SB221 (previously called JA221 lpp^-, see NAKAMURA and INOUYE 1979) containing pINIII-A-nuclease (no signal control), pINIII-B-nuclease (nuclease hybrid with *lpp* signal), pINIII-C-nuclease (nuclease hybrid with *lpp* signal plus nine amino acids), and pINIII-*ompA*-nuclease (nuclease hybrid with *ompA* signal) were induced for 2 h with 2 mM isopropyl-β-D-thio-galactopyranoside (IPTG). Extracts prepared from these cultures were then analyzed by sodium dodecyl sulfate (SDS) polyacrylamide gel electrophoresis. Figure 4 shows that the maximum amount of nuclease is produced by the pINIII-*ompA*-nuclease (lane 10). The gene product of pINIII-B-nuclease (lane 7) migrated significantly slower than the pINIII-*ompA*-nuclease gene product. Pulse-chase experiments (data not shown) suggest that the signal peptide was not cleaved from the pINIII-B-nuclease gene product. The gene product of pINIII-C-nuclease (lane 8) also migrated significantly slower than the pINIII-ompA-nuclease product. This gene product was labeled with palmitic acid, and the species was shifted to a higher molecular weight with the addition of globomycin, an inhibitor of the lipoprotein specific signal peptidase II (HUSSAIN et al. 1980). This suggests that the pINIII-C-nuclease is modified, and is cleaved by signal peptidase II, as is observed when β-lactamase is cloned into pINIII-C (GHRAYEB and INOUYE 1984). The higher molecular weight is due to the presence of additional amino acids derived from the linker sequences (see Fig. 3) and the glyceride modification. The linker sequence can be removed by oligonucleotide-directed site-specific mutagenesis (MASUI et al. 1984). The amount of nuclease product was determined on Coomassie blue stained gels by densitometry. Table 2 confirms that nuclease cloned into the pINIII-ompA vector produces the most nuclease, representing 9% of the total cellular protein. This value corresponds to approximately 300 mg nuclease/liter of culture (see Table 1). Cloning into the pINIII-C, pINIII-B, and pINIII-A vectors produced nuclease to 6%, 2%, and 1% of total cell protein, respectively.

We next determined the localization of the cloned gene products. Crude membrane fractions were first prepared from extracts of induced cells harboring

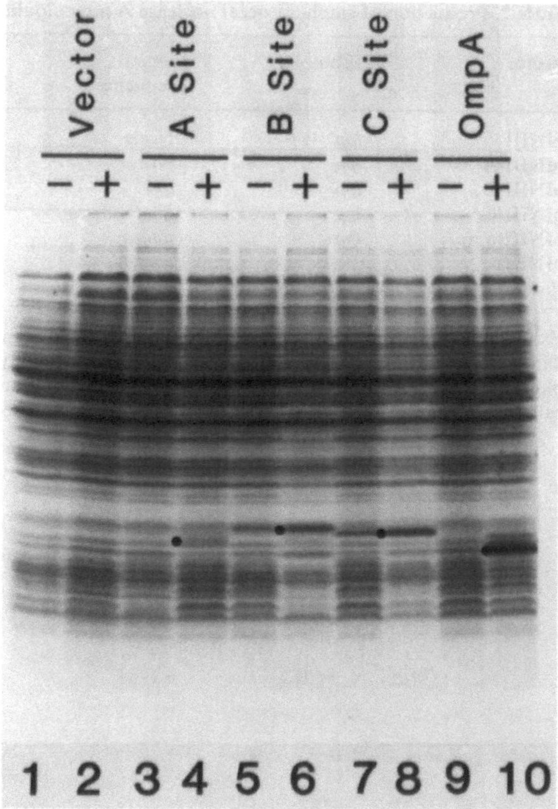

Fig. 4. Production of staphylococcal nuclease A cloned into secretion cloning vectors. Exponentially growing *E. coli* SB221 containing pINIII (lanes 1 and 2), pINIII-A-nuclease (lanes 3 and 4), pINIII-B-nuclease (lanes 5 and 6), pINIII-C-nuclease (lanes 7 and 8), or pINIII-ompA-nuclease (lanes 9 and 10) was induced for 2 h with 2 m*M* IPTG (*even-numbered lanes*). Extracts prepared from these cultures were analyzed by SDS polyacrylamide gel electrophoresis, followed by staining with Coomassie brilliant blue

each plasmid. The membranes were then fractionated into inner membrane and outer membrane fractions by differential solubility in *N*-lauroyl sarcosinate (Filip et al. 1973). As expected, the gene product of pINIII-A-nuclease was found in the cytoplasm. The unprocessed gene product of pINIII-B-nuclease was found associated with the cytoplasmic membrane, based upon differential solubility of the membrane-associated polypeptide with sodium sarcosinate. Further studies showed that the membrane-bound nuclease was digested by proteases added to a suspension of spheroplasts. These results suggest that pINIII-B-nuclease produces a form of nuclease that interacts with the outer surface of the cytoplasmic membrane, presumably via the hydrophobic signal peptide. The processed, modified gene product of pINIII-C-nuclease associated with the outer membrane fraction, again based on differential solubility with sodium sarcosinate. This result is consistent with data from pINIII-C-β-lactamase (Ghrayeb and Inouye 1984). Finally, the processed product of pINIII-ompA-nuclease was released from cells by osmotic shock, a result consistent with its being localized in the periplasmic space.

With this collection of vectors, one can position a secreted protein in any of four compartments of the bacterial cell: pINIII-A for cytoplasmic localiza-

Table 2. Production of staphylococcal nuclease A after cloning into secretion cloning vectors

Vector	Promoter	Signal sequence	Protein	Production (%)
pINIII	*lpp*	None	None	0
pINIII-A	*lpp*	None	Nuclease	1
pINIII-B	*lpp*	*lpp*	Nuclease	2
pINIII-C	*lpp*	*lpp* (+8 aa)	Nuclease	6
pINIII-ompA	*lpp*	*ompA*	Nuclease	9
pINIII-(*lpp*^{P}-5)-ompA	*lpp*^{P}-5	*ompA*	Nuclease	42

Exponentially growing *E. coli* SB221 containing the designated vector was induced for 2 h with 2 m*M* IPTG. Extracts prepared from these cultures were analyzed by SDS polyacrylamide gel electrophoresis, stained with Coomassie brilliant blue, and the amount of protein quantitated densitometrically. Data from pINIII(*lpp*^{P}-5)-*ompA* nuclease were obtained from INOUYE and INOUYE (1985)

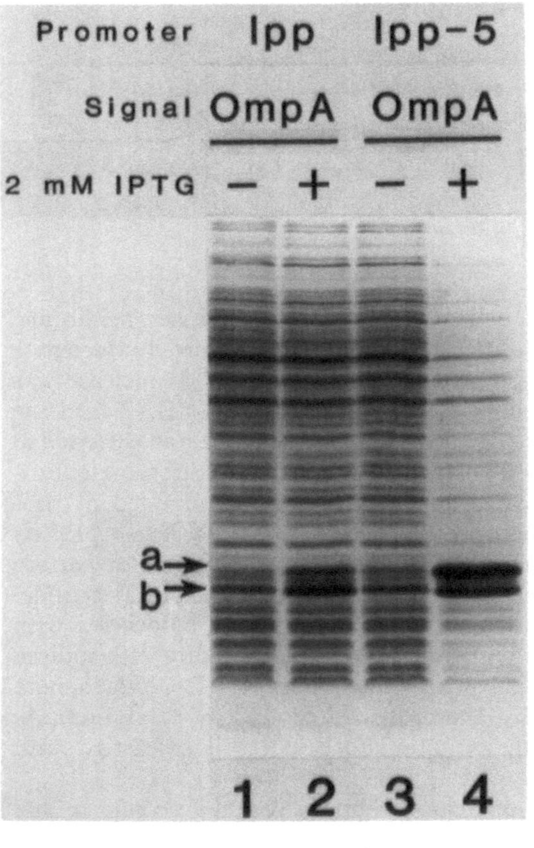

Fig. 5a, b. Production of staphylococcal nuclease A cloned into secretion cloning vector utilizing optimized lipoprotein promoter sequence. Exponentially growing *E. coli* SB221 containing pINIII-*ompA* nuclease (lanes 1 and 2) or pIN(*lpp*^{P}-5)-*ompA* nuclease (lanes 3 and 4) was induced for 3 h with 2 m*M* IPTG (lanes 2 and 4). Whole cell extracts prepared from these cultures were analyzed by SDS polyacrylamide gel electrophoresis, followed by staining with Coomassie brilliant blue. *Arrows a and b* correspond to unprocessed (**a**) and processed (**b**) *ompA* nuclease

tion, pINIII-B for localization in the cytoplasmic membrane, pINIII-ompA for localization in the periplasm, and pINIII-C for localization within the outer membrane. Based on total protein production (Table 2), the pINIII-ompA vector seemed the optimal secretion vector for the production of large amounts of native protein in *E. coli*. We have sought to improve this vector further by optimizing its lipoprotein promoter. Using oligonucleotide-directed site-specific mutagenesis, we found (Inouye and Inouye 1985) that by changing the −35 region of the *lpp* promoter from TTCTCA to TTGACA, and by changing the −10 region from AATACT to TATACT, expression of β-galactosidase could be increased four fold. We therefore incorporated this altered *lpp* promoter, called $lpp^{p}-5$, into the pINIII-*ompA* secretion cloning vector, and compared its production of nuclease with that of pINIII-*ompA* nuclease. Figure 5 shows that the pINIII(lpp^{p}-5)-*ompA*-nuclease (lane 4) produced much more gene product than the pINIII-*ompA* nuclease (lane 2). Quantitation of the Coomassie-stained gel by densitometry showed that the pINIII(lpp^{p}-5)-ompA-nuclease gene product corresponds to roughly 50% of the total cellular protein (Table 2). This represents approximately 1.5 g nuclease produced/liter culture. Figure 5 also shows that only a portion (40%) of the nuclease produced by pINIII(*lpp*^{p}-5)-*ompA*-nuclease was processed by the signal peptidase.

6 Discussion

As described, secretion cloning vectors present numerous advantages in the production of cloned gene products in bacteria. For cloning of cytotoxic agents, for expediting native protein purification, and for other reasons described previously, secretion cloning vehicles may be the vehicle of choice for some experiments. As these cloning systems are used more widely, we will accumulate more detailed information of the rules governing the use of these vectors. For example, the presence of stop-transfer sequences [a hydrophobic core peptide flanked by charged residues (Rose et al. 1980; Yost et al. 1983) can inhibit translocation of β-galactosidase (Silhavy et al. 1979; Fowler and Zobin 1978). It may be possible, by inserting a charged residue within the central hydrophobic domain in the stop-transfer sequence, that this block can be removed without significant change in the protein's enzymatic activity. Such mutations are known to alter signal peptide function in lambda receptor protein (Emr et al. 1980), maltose-binding protein (Bedouelle et al. 1980), alkaline phosphatase (Michaelis et al. 1983), and lipoprotein (Lin et al. 1980). The use of oligonucleotide-directed site-specific mutagenesis (Masui et al. 1984) makes such experiments almost routine. Such a strategy has been successfully employed to inhibit secretion of lipoprotein (Pollitt et al. 1985).

Further study will continue to elucidate the benefits choosing either *Bacillus subtilis* or *Escherichia coli* as bacterial host for protein production. At present, *E. coli* seems the organism of choice. This bacterium has few extracellular proteases to damage the cloned gene product. Our knowledge of *E. coli* genetics has allowed production of extremely large amounts of the cloned gene product

(Tables 1, 2). Finally the possibility of generating gram-negative bacteria with leaky outer membranes by mutation of outer membrane synthetic enzymes, or by introducing inducible glycosylases into the cell, may further simplify protein purification. Regardless, further study on the control of protease biosynthesis in *B. subtilis* may lead to the perfect organism for large-scale protein production.

Further improvements of the cloning vehicle itself may be possible. We have shown that optimizing promoter efficiency in the *E. coli* pINIII-ompA cloning vector resulted in cells whose nuclease content constitutes 42% of the total cellular protein (Fig. 5, Table 1). This level of production saturated the protein translocation pathway, as only 40% of the cloned gene product was processed. By determining the step of translocation limited in vivo, it may be possible to complement the deficiency by cloning the appropriate gene into the secretion cloning vector. For example, the accumulation of nuclease precursor may result from insufficient signal peptidase I, a protein present in 500 copies per cell (Wolfe et al. 1982). Assuming the cloned signal peptidase can correctly insert into the cytoplasmic membrane, cloning the gene into the cloning vector may improve processing efficiency of the foreign gene products. Such an improvement in secretion cloning vectors promises to increase the usefulness of these vehicles in production of proteins in *E. coli.*

Acknowledgments. We would like to thank D. Comeau and Dr. P. March for their critical reading of this manuscript. This work was supported by grants from the National Institutes of Health (GM 19043), the American Cancer Society (Np 3871), and the Eli Lilly Company.

References

Bedouelle H, Bassford PJ, Fowler AV, Zabin I, Beckwith J, Hofnung M (1980) Mechanisms of protein localization. Nature 285:78–81

Blobel G (1980) Intracellular protein topogenesis. Proc Natl Acad Sci USA 77:1496–1500

Chang S, Gray O, Ho D, Kroyer J, Chang S-Y, McLaughlin J, Mark D (1981) Expression of eukaryotic genes in B. subtilis using signals of penP. In: Hoch JA, Ganesan AT, Chang S (eds) Molecular cloning and gene regulation in Bacilli. Academic, New York, pp 159–169

Chang S, Ho D, Gray O, Chang S-Y, McLaughlin J (1983) Functional expression of human interferon genes and construction of a partition-proficient plasmic vector in *B. subtilis*. In: Ikeda Y, Beppu T (eds) Genetics of industrial microorganisms. Kodansha, Tokyo, pp 227–231

Duffaud GD, Lehnhardt SK, March PE, Inouye M (1985) Structure and functions of the signal peptide. In: Knauf PA, Cook JS (eds) Current topics in membranes and transport, vol 24. Academic, London, pp 65–104

Emr S, Hedgpeth J, Clement J, Silhavy TJ, Hofnung M (1980) Sequence analysis of mutations that prevent export of λ receptor, an Escherichia coli outer membrane protein. Nature (London) 285:82–85

Filip C, Fletcher G, Wulff JL, Earhart CF (1973) Solubilization of the cytoplasmic membrane of *Escherichia coli* by the ionic detergent sodium-lauryl sarcosinate. J Bacteriol 115:717–722

Fowler AV, Zabin I (1978) Amino acid sequence of β-galactosidase. XI. Peptide ordering procedure and the complete sequence. J Biol Chem 253:5521–5525

Ghrayeb J, Inouye M (1984) Nine amino acid residues at the amino terminal of lipoprotein are sufficient for its modification, processing and localization in the outer membrane of *Escherichia coli.* J Biol Chem 259:463–467

Ghrayeb J, Kimura H, Takahara M, Hsiung H, Masui Y, Inouye M (1984) Secretion cloning vectors in *Escherichia coli.* EMBO 3:2437–2442

Gould AR, May BK, Elliot WH (1975) Release of extracellular enzymes from *Bacillus amyloliquefaciens.* J Bacteriol 122:34–40

Hussain M, Ichihara S, Mizushima S (1980) Accumulation of glyceride-containing precursor of the outer membrane lipoprotein in the cytoplasmic membrane of *Escherichia coli* treated with globomycin. J Biol Chem 255:3707–3712

Inouye M, Halegoua S (1980) Secretion and membrane localization of proteins in *Escherichia coli.* CRC Crit Rev Biochem 7:339–371

Inouye S, Inouye M (1985) Up-promoter mutations in the *lpp* gene of *Escherichia coli.* Nucleic Acids Res 13:3101–3110

Inouye M, Nakamura K, Inouye S, Masui Y (1983) Versatile expression cloning vehicles using the lipoprotein gene of the *Escherichia coli* and their application. In: Watanabe I (ed) The Future in Nucleic Acid Research. Academic Japan, Tokyo, pp 419–436

Lin JJC, Kanazawa H, Wu HC (1980) Purification and characterization of the outer membrane lipoprotein from an *Escherichia coli* mutant altered in the signal sequence of prolipoprotein. J Biol Chem 255:1160–1163

Lundstrom K (1984) Expression of the vesicular stomatitis membrane glycoprotein gene in *Bacillus subtilis.* FEMS Microbiol Letters 23:65–70

Masui Y, Coleman J, Inouye M (1983) Multipurpose expression cloning vehicles in *E. coli.* In: Inouye M (ed) Experimental manipulation of gene expression. Academic, New York, pp 15–32

Masui Y, Mizuno T, Inouye M (1984) Novel high-level expression cloning vehicles: 10^4-fold amplification of *Escherichia coli* minor protein. Bio/Technology 2:81–85

Michaelis S, Inouye H, Oliver D, Beckwith J (1983) Mutations that alter the signal sequence of alkaline phosphatase in *Escherichia coli.* J Bacteriol 154:366–374

Moreno F, Fowler AV, Hall M, Silhavy TJ, Zabin I, Schwartz M (1980) A signal sequence is not sufficient to lead β-lactamase out of the cytoplasm. Nature 286:356–359

Morinaga Y, Franceschini T, Inouye S, Inouye M (1984) Improvement of oligonucleotide-directed site-specific mutagenesis using double-stranded plasmid DNA. Bio/Technology 2:636–639

Nakamura K, Inouye M (1979) DNA sequence of the gene for the outer membrane lipoprotein of *E. coli:* An extremely AT-rich promoter. Cell 18:1109–1117

Ohmura K, Shiroza T, Nakamura K, Nakayama Y, Yamane K, Yoda K, Yamasaki M, Tamura G (1984) A *Bacillus subtilis* secretion vector system derived from the *B. subtilis* α-amylase promoter and signal sequence region, and secretion of *Escherichia coli* β-lactamase by the vector system. J Biochem 95:87–93

Palva I, Sarvas M, Lehtovaara P, Sibakov M, Kaariainen L (1982) Secretion of *Escherichia coli* β-lactamase from *Bacillus subtilis* by the aid of α-amylase signal sequence. Proc Natl Acad Sci USA 79:5582–5586

Palva J, Lehtovaara P, Kaariainen L, Sibakov M, Cantell K, Schein CH, Kashiwagi K, Weissmann C (1983) Secretion of interferon by *Bacillus subtilis.* Gene 22:229–235

Pollitt S, Inouye S, Inouye M (1985) Functional analysis of signal peptide mutants. In: Leive L, Bonventre PF, Morello JA, Schlesinger S, Silver SD, Wu HC (eds) Microbiology 1985. American Society for Microbiology, Washington, DC, pp 308–311

Randall LL, Hardy SJS (1984) Export of protein in bacteria: Dogma and data. Modern Cell Biol 3:1–20

Rose JK, Welch WJ, Sefton BM, Esch FS, Ling NC (1980) Vesicular stomatitis virus glycoprotein is anchored in the viral membrane by a hydrophobic domain near the COOH terminus. Proc Natl Acad Sci USA 77:3884–3888

Schultz GE, Schirmer RH (1979) Principles of protein structure. Springer, Berlin Heidelberg New York, p 39 (Springer advanced cyts in chemistry)

Shine J, Dalgarno L (1975) Determination of cistron specificity in bacterial ribosomes. Nature 254:34–38

Shiroza T, Nakazawa K, Tashiro N, Yamane K, Yanagi K, Yamasaki M, Saito H, Kawade Y, Taniguchi T (1985) Synthesis and secretion of biologically active mouse interferon-β using a *Bacillus subtilis* α-amylase secretion vector. Gene 34:1–8

Sibakov M, Lehtovaara P, Pettersson R, Lundstrom K, Kaalkkinen N, Ulmanen I, Takkinen K, Kaariainen L, Palva I, Sarvas M (1984) Secretion of foreign gene products by the aid of a

Bacillus secretion vector. In: Ganesan AT, Hoch JA (eds) Genetics and biotechnology of bacilli. Academic, New York, pp 153–162

Silhavy TJ, Bassford PJ, Beckwith JR (1979) A genetic approach to the study of protein translocation in *Escherichia coli*. In: Inouye M (ed) Bacterial outer membranes: Biogenesis and functions. Wiley, New York, pp 203–254

Silhavy T, Benson SA, Emr SD (1983) Mechanisms of protein localization. Microbiol Rev 47:313–344

Stueber D, Bujard H (1982) Transcription from efficient promoters can interfere with plasmid replication and diminish expression of plasmid specified genes. EMBO J 1:1399–1404

Takahara M, Hibler DW, Barr PJ, Gerlt JA, Inouye M (1985) OmpA signal peptide directed secretion of staphylococcal nuclease A by *Escherichia coli*. J Biol Chem 260:2670–2674

Tommassen J, van Tol H, Lugtenberg B (1983) The ultimate localization of an outer membrane protein of *Escherichia coli* K12 is not determined by the signal sequence. EMBO 2:1275–1279

Ulmanen I, Lundstrom K, Lehtovaara P, Sarvas M, Ruohonen M, Palva I (1985) Transcription and translation of foreign genes in *Bacillus subtilis* by the aid of a secretion vector. J Bacteriol 162:176–182

von Heijne G (1980) Trans-membrane translocation of proteins: A detailed physico-chemical analysis. Eur J Biochem 103:431–438

Wolfe PB, Silver P, Wickner W (1982) The isolation of homogeneous leader peptidase from a strain of *Escherichia coli* which overproduces the enzyme. J Biol Chem 257:7898–7902

Yost CS, Hedgpeth J, Lingappa VR (1983) A stop transfer sequence confers predictable transmembrane orientation to a previously secreted protein in cell-free system. Cell 34:759–766

Signal Peptidases

P. Ray[1], I. Dev[1], C. MacGregor[2], and P. Bassford, Jr.[3]

1 Introduction

This review will attempt to summarize the current literature concerned with the in vitro processing of precursor proteins by solubilized signal peptidases; by necessity, it will mainly be concerned with the bacterial signal peptidases. A number of recent reviews concerned with various aspects of protein export have been published (Benson et al. 1985; Bankaitis et al. 1985; Randall and Hardy 1984; Silhavy et al. 1983) and these topics will not be discussed, except in the context that they pertain to the signal peptidases.

[1] Department of Microbiology, Wellcome Research Laboratories, 3030 Cornwallis Road, Research Triangle Park, NC 27709, USA
[2] Department of Microbiology, Georgetown University Medical School, Washington, DC 20007, USA
[3] Department of Microbiology and Immunology, University of North Carolina, School of Medicine, Chapel Hill, NC 27514, USA

Current Topics in Microbiology and Immunology, Vol. 125

Signal peptidases are responsible for the specific removal of a small polypeptide, designated the signal peptide or the leader peptide, from the amino terminus of most secretory proteins. According to the signal hypothesis, the role of the signal peptide in eukaryotic cells would be to initiate the vectorial transfer of the secreted protein from the cytoplasm, the site of protein synthesis, through the membrane and into the lumen of the endoplasmic reticulum. The removal of the signal peptide by the membrane-bound signal peptidase serves to release the secretory protein from the membrane and make the transfer an irreversible event.

In *Escherichia coli*, the major bacterial system utilized to study protein export, the role of the signal peptidases in protein translocation is being investigated in a number of laboratories. In gram-negative cells, proteins that exist on the outer surface of the cytoplasmic membrane, on or in the outer membrane, or in the periplasmic space can be considered secretory proteins. As such, most of these proteins have signal peptides that are necessary for initiating the export event and that are cleaved by the signal peptidases during or shortly after translocation across the cytoplasmic membrane. The data clearly demonstrate that cleavage of the signal peptide is not required for translocation; however, translocation is required for cleavage (SILHAVY et al. 1983; RANDALL and HARDY 1984; BENSON et al. 1985). There are three questions concerning the role of the signal peptidases in protein translocation that have recently been partially answered: (1) Is processing required for the proper localization of a particular protein? (2) Is processing required for the activity or function of a particular protein? (3) Is processing by the signal peptidases a required function or, in other words, is the signal peptidase an essential enzyme?

From both biochemical and genetic data obtained using *E. coli*, the answer to the first question appears to be yes with some qualifications; in order for the secreted proteins to be localized correctly, the precursor forms of most proteins must be processed. Mizushima's laboratory (HUSSAIN et al. 1980, 1982a) has shown that the modified precursor of Braun's lipoprotein, which normally exists in the outer membrane, accumulates in the cytoplasmic membrane in the presence of globomycin, a specific inhibitor of the prolipoprotein signal peptidase (INUKAI et al. 1978; HUSSAIN et al. 1980); some of the precursor is covalently bound to the peptidoglycan. Even though this precursor was not solubilized by sodium sarcosinate (INUKAI and INOUYE 1983), suggesting that the protein is also in association with the outer membrane, the important data show clearly that the amino terminal portion of the signal peptide is within the cytoplasm. If the globomycin is diluted out, then the precursor is processed and the mature protein is correctly localized. DALBEY and WICKNER (1985) have recently shown that in the absence of functional SPase I the precursors of OmpA, an outer membrane protein, and MBP (maltose-binding protein), a periplasmic protein, are translocated across the cytoplasmic membrane but remain anchored to the outer surface of the inner membrane. While studying the export of TEM-1-β lactamase, KOSHLAND et al. (1982) showed that processing was required for the correct localization of this protein to the periplasmic space. However, with certain mutations in the signal sequence region of the genes encoding for LamB and MBP, correct localization can be achieved without processing (Silhavy et al. 1983; J. Fikes and P. Bassford, unpublished). The original signal sequence mutation isolated by Lin et al. (1978, 1980a, b), in which the ninth amino acid in the lipoprotein signal sequence was converted from glycine to aspartic acid, blocks processing and modification but not localization. However, the amount of lipoprotein synthesized in this mutant is reduced, the rate of translocation is decreased, and not all of the lipoprotein is correctly localized. These data suggest that processing is required for the proper localization of most proteins to the outer membrane and to the periplasm but is not obligatory in all situations; some precursors can be correctly localized at a slower rate.

The second question concerning the requirement for processing for the activity or function of the exported protein seems to depend on the protein in question. The initial data of Inouye and Beckwith (1977) showed that the precursor of alkaline phosphatase synthesized in an in vitro transcription translation system when dialyzed in the presence of Zn^{2+} exhibited phosphatase activity. The precursor of MBP, also synthesized in vitro, was shown to bind its substrates, maltose, and higher maltodextrins (Ferenci and Randall 1979). Mutations that prevent the processing but not the export of LamB and β-lactamase do not inhibit function or enzyme activity (Emr and Silhavy 1980; Koshland et al. 1982). Even though some proteins have been shown to be functional, other proteins have been shown to be nonfunctional in their precursor form. When the structural gene for the F sex factor outer membrane surface exclusion protein (*traT*) was cloned onto a high-copy number plasmid (Minkley 1984), the precursor protein was translocated to the outer membrane; however, it was defective in forming the oligomeric surface structure required for the exclusion function. The precursor of LamB synthesized in vitro failed to inactivate phage lambda (Marchal et al. 1980), suggesting that the precursor could not form an oligomer; however, as noted above, certain mutations of LamB that are not processed are localized and do bind lambda phage (Emr and Silhavy 1982). Recently, Roggenkamp et al. (1981, 1983) have shown that the ampicillin resistance gene (*bla*) from *E. coli* could be expressed in *Saccharomyces cerevisiae* and was active in the processed form. However, the precursor form synthesized in yeast was inactive (Roggenkamp et al. 1985). These data indicate that processing may or may not be necessary for the activity or function of a specific protein; however, processing per se may not be required for activity or function but processing may be required for correct folding (compare the data for LamB synthesized in vivo and in vitro).

From both biochemical and genetic studies, the answer to the third question concerned with the necessity of the signal peptidases is clear. Globomycin is an antibiotic that specifically inhibits the proliprotein signal peptidase. When cells are grown in the presence of globomycin, the precursor prolipoprotein accumulates and cell growth is inhibited (Hussain et al. 1980; Ichihara et al. 1982). Yamagata et al. (1983a) have isolated a temperature-sensitive mutant that maps in the *lsp* gene that at elevated temperatures causes the accumulation of prolipoprotein. These data indicate that SPase II (signal peptidase II or prolipoprotein signal peptidase) is an essential enzyme required for normal bacterial growth. SPase I (signal peptidase I or leader peptidase) has been shown to be an essential enzyme by DATE (1983) and by DALBEY and WICKNER (1985) using two different genetic techniques. DATE (1983) showed that the inactivation of the SPase I gene (*lep*) was a lethal event. Subsequently, DALBEY and WICKNER (1985) constructed an *E. coli* strain in which expression of the *lep* gene is under the control of the arabinose promoter. In the presence of arabinose, the cells grew normally; however, in the absence of arabinose, when the *lep* gene product was no longer produced, cell growth ceased and the precursors of at least two proteins, OmpA and MBP, were accumulated on the outer surface of the inner membrane. These results show that two signal peptidases isolated from *E. coli* are essential for normal cellular growth; the inhibition of function or the lack of synthesis of the signal peptidase results in the accumulation of precursors on the outer surface of the inner membrane.

2 Preparation of Substrates

The substrates utilized for the in vitro assay of the various signal peptidases are precursor proteins that must be synthesized either in vitro or in vivo. To our knowledge, there is no proven chromogenic synthetic peptide (see STRAUSS et al. 1979) that serves as a substrate for these specific enzymes. There are a number of methods by which precursor proteins can be prepared and purified; however, as pointed out by JACKSON (1983), not all full-length precursor proteins are processed with equal efficiency. It appears, particularly in the eukaryotic systems studied, that not all of the precursors can be processed to the mature form. Since in the eukaryotic system, the translocation and processing event is cotranslational, this may not be unexpected. The full-length precursor protein, utilized in in vitro post-translational processing reactions, might fold and thus obscure the processing site. There is evidence in the bacterial system which suggests that the protein precursors made in vitro become translocationally incompetent in a time-dependent manner (CHEN et al. 1985; GOODMAN et al. 1981). These data also suggest that the precursors may change conformation and block recognition of the processing site.

2.1 In Vitro Translation from DNA or RNA

The most commonly used substrates for the mammalian SPase are either pre-human placental lactogen (p-hPL, STRAUSS et al. 1979; LIVELY and WALSH 1983) or bovine preprolactin (p-PL, JACKSON and BLOBEL 1977; STERN and JACKSON 1985). These substrates are made in cell-free protein-synthesizing systems using either a wheat germ or rabbit reticulocyte lysate programmed with the poly(A)-containing RNA isolated from bovine pituitaries or fresh human placentae. Stocks of the [^{35}S]-methionine-labeled translation products are stored at $-80°$ C until utilized. No specific activities are available on the labeled precursors.

Precursors of the bacterial proteins have been synthesized using either plasmid DNA or isolated RNA to direct the synthesis of the precursor proteins in cell-free protein-synthesizing systems. WICKNER and coworkers (WICKNER et al. 1978; SILVER et al. 1981) have utilized either M13 duplex DNA or mRNA isolated from M13-infected cells to program the synthesis of [^{35}S]-methionine-labeled procoat, in which 30%–90% of the label is found in the precursor protein. No specific activities are available on the labeled precursors.

2.2 Membrane Perturbants and Inhibitors

A number of radiolabeled precursors have been detected in bacterial cells treated with agents that affect membrane fluidity or energy metabolism such as phenylethyl alcohol (HALEGOUA and INOUYE 1979), procaine (LAZDUNSKI et al. 1979), or carbonyl cyanide *m*-chlorophenylhydrazone (DANIELS et al. 1981). If the precursor proteins are labeled in the presence of these agents, the specific labeled proteins can be isolated and utilized as substrates for the signal peptidases if antibody is available (see Sect. 2.4; LAZDUNSKI et al. 1979). In our hands, the above agents are not dependable since the yield of precursor to mature protein is variable, the labeling in the presence of these agents is reduced, and the membrane perturbants do not cause general precursor accumulation (SILHAVY et al. 1983).

The substrate for SPase II, the modified prolipoprotein, has been prepared in vivo using *E. coli* B grown in the presence of globomycin. Globomycin, or SF1902-A1, is a cyclic peptide antibiotic that inhibits the growth of *E. coli* and causes the modified form of prolipoprotein to accumulate in the cytoplasmic membrane (INUKAI et al. 1978; OMOTO et al. 1981; HUSSAIN et al. 1980). According to the method of DEV and RAY (1984), substrate amounts of the prolipoprotein can be prepared in the presence of globomycin. A typical preparation of the purified radiolabeled prolipoprotein substrate is better than 90% pure, as determined by radioactivity; however, it is contaminated with about 40% unlabeled lipoprotein (the presence of the unlabeled mature protein does not affect processing). The specific radioactivity of the prolipoprotein prepared in this manner is approximately 5.5×10^5 counts per minute per nanomole. Globomycin has been utilized to prepare the prolipoprotein substrate in the laboratories of MIZUSHIMA and WU (HUSSAIN et al. 1980, 1982a; YAMAGATA et al. 1983a; TOKUNAGA et al. 1982b; TOKUNAGA and WU 1984), who have utilized this substrate in activity assays to clone SPase II and to show that SPase II is different from SPase I (TOKUNAGA et al. 1982a).

2.3 Strains Containing Fusion Proteins

BECKWITH and coworkers (BASSFORD et al. 1979; SILHAVY et al. 1977, 1979) utilized the technique of gene fusion (CASADABAN 1976) to construct *E. coli* strains producing hybrid proteins having the amino terminal sequence of an exported protein (e.g., MBP, LamB) preceding an enzymatically active *β*-galactosidase moiety. If the hybrid protein includes an intact amino terminal signal sequence then, upon induction of hybrid gene expression, export of the hybrid protein is initiated in a manner probably analogous to the native protein. However, due to its carboxy-terminal β-galactosidase moiety, the hybrid protein cannot traverse the cytoplasmic membrane, becomes lodged there, and somehow blocks the export of virtually all signal peptide-bearing proteins. These proteins then accumulate in the cytoplasm as unprocessed precursors. It has been suggested that the hybrid protein progressively occupies specific sites in the membrane required for general protein export (BASSFORD et al. 1979; ITO et al. 1981). TOKUNAGA et al. (1982b) and DEV and RAY (1984) have utilized an *E. coli* strain containing a specific MalE-LacZ gene fusion to generate the unmodified precursor of prolipoprotein. The same strain was also utilized to isolate radiolabeled precursors of MBP (ITO 1982; RAY et al. to be published) as well as PhoA and LamB using specific antibody coupled to Sepharose (P.H. Ray and I.K. Dev, unpublished results). The use of the gene fusion strains to prepare specific radiolabeled precursor proteins is feasible if specific antibody or affinity ligand to the protein is available; however, this is not the method of choice since the ratio of radiolabeled precursor to mature is variable (there is variation in the time of induction before translocation is blocked and there is some translocation, thus processing, even after 240 min after induction) and the actual yield of labeled precursor is low (P.H. Ray and I.K. Dev, unpublished results).

2.4 Strains Containing Signal Sequence Mutations

Certain mutational alterations within the signal sequence of MBP, LamB, Bla, and PhoA interfere with the export process and allow the various precursors to accumulate within the cytoplasm in their unprocessed form (summarized in BANKAITIS et al. 1985; BENSON et al. 1985; MICHAELIS and BECKWITH 1982). Any signal sequence mutation that allows the precursor protein to accumulate within the cytoplasm is a good source for the isolation of high specific activity radiolabeled precursor proteins that can be utilized as substrates for the signal peptidases. Specifically, we have utilized a number of signal sequence mutations in the *malE* gene to produce substrates used in the isolation of another signal peptidase and to assay SPase I. The amino acid sequence of the wild-type MBP signal peptide and the various signal sequence mutants used in these studies are shown in Fig. 1.

The procedure to isolate the precursors of MBP derived from the above signal sequence mutants is as follows. The strains are grown overnight in 0.2% maltose minimal medium, diluted to an optical density at 600 nm (OD 600) of 0.05, allowed to grow at 30° C to an OD of 0.3, and then 50-ml cultures are labeled with 1–5 mC of [^{35}S]-methionine for 1.5–3 min, depending upon the strength of the mutation (EMR and BASSFORD 1982). Phenylethyl alcohol (0.3%) is then added for 5 min and the labeling is terminated with carbonyl cyanide m-chlorophenylhydrazone (CCCP) (30 µ*M*) and an equal volume of ice. The culture is centrifuged, the pellet is suspended in 100–400 µl 1% sodium dodecyl sulfate (SDS), containing 50 m*M* Tris-Cl pH 6.8 and 1 m*M* ethylenediaminetetraacetate (EDTA), and boiled for 5 min. The sample, diluted with 1% Triton X-100 antibody buffer (ITO et al. 1981), is centrifuged and the supernatant

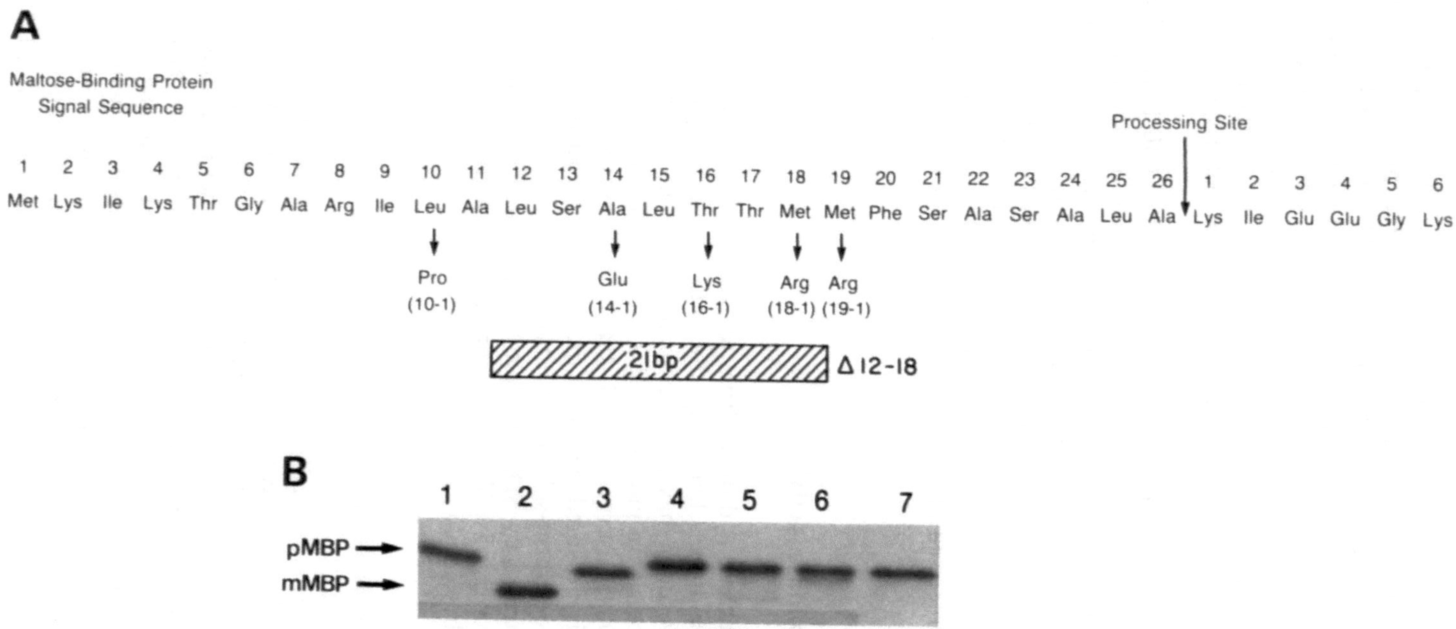

Fig. 1. A The amino acid sequence of the first 32 amino acids of the pre-maltose-binding protein including the signal sequence and the processing site. Single amino acid substitutions resulting in an export defect are indicated by *arrows*; the 21-bp deletion is indicated by *shaded bars*. **B** Autoradiogram of the precursors generated from the above signal sequence mutants. The precursors were labeled, isolated, and analyzed by SDS-gel electrophoresis as described in the text. Lanes 1, 3, 4, 5, 6, and 7 contain aliquots of the precursors obtained from mutants 10-1, Δ12–18, 14-1, 16-1, 18-1, and 19-1 respectively. Lane 2 contains mature MBP isolated from strain MC4100

bound to antibody coupled to Sepharose and washed with antibody buffer. The labeled MBP precursor is eluted with 0.2 *M* HCl-glycine, pH 2.3 the peak radioactive fractions are pooled, dialyzed against 0.01 *M* $(NH_4)HCO_3$, and lyophilyzed. The precursors of MBP prepared in this manner have a specific activity of $1–3 \times 10^7$ counts per minute per nanomole. The wild-type precursor is labeled and isolated in the same way from a strain containing a MalE-LacZ gene fusion after induction with 0.3% maltose for 240 min. Under these conditions, the labeled precursor represents 90% of the total radiolabeled MBP recovered.

Recently, the wild-type precursor was isolated from the MalE-LacZ gene fusion strain in the absence of SDS. The cells were labeled and the reaction was terminated with CCCP and ice. After centrifugation, the cells were suspended in 50 m*M* Tris-Cl containing 10 µg/ml lysozyme and frozen at −20° C. The cells were thawed and frozen twice, then diluted 1:20 in 1% Triton X-100 antibody buffer. After centrifugation, the supernatant was bound to the antibody-Sepharose column as before. The substrates prepared in the absence and presence of SDS were processed at the same rate with SPase I and SPase III (see Sect. 5).

3 Assay of Signal Peptidase Activity

The assay of signal peptidases is difficult from the kinetic point of view, since very few of the substrates have a known specific activity and the quantitation of the product of the reaction has to be estimated from an autoradiogram after SDS-gel electrophoresis. In most cases (Jackson and Blobel 1977; Zwizinski and Wickner 1980; Tokunaga et al. 1982a, b; Hussain et al. 1980), signal peptidase activity is measured in vitro by incubating the labeled full-length precursor protein with a detergent-solubilized enzyme preparation. The products of the reaction are separated by SDS-gel electrophoresis and estimated after autoradiography or fluorography. The precursor proteins migrate as a slower migrating species since they contain the signal peptide.

3.1 SDS-Gel Electrophoresis Assay

Signal peptidase activity was originally detected by Blobel and Dobberstein (1975) in a cotranslational assay measuring the translocation into microsomal membrane vesicles by the protection of the mature product from degradation by externally added proteases. Subsequently, post-translational processing was effected by incubating radiolabeled precursors with detergent extracts and terminating the reactions with either trichloroacetic acid or SDS sample buffer. The amount of mature product produced from the precursor is quantitated after autoradiography or fluorography by visual examination, by excising the radioactive band from the gel and counting in a scintillation counter or by scanning the autoradiogram with a laser densitometer (Tokunaga et al. 1982a, b; ZWIZINSKI and WICKNER 1980; JACKSON 1983). The preferred method of stopping the reaction is the addition of SDS-sample buffer. In our hands, precipitation of the reaction with 5% trichloroacetic acid (TCA) and washing the pellet with acetone to remove the TCA sometimes leads to a pellet that is partially insoluble in the SDS sample buffer, particularly in those cases when a reticulocyte lysate translation system is utilized.

Mammalian SPases are usually assayed utilizing radiolabeled pre-hPL or pre-PL. A typical reaction mixture (STERN and JACKSON 1985) contains 5 μl radiolabeled pre-PL and 20 μl microsomal detergent extract supplemented with 114 nmol soybean phospholipids. The reaction for the mammalian enzyme is carried out for either 30 or 90 min at 25° C and is terminated with either 2X sample buffer or 66% cold acetone (JACKSON and BLOBEL 1977; STRAUSS et al. 1979; STERN and JACKSON 1985; FUJIMOTO et al. 1984). The specific activities of the substrates have not been determined and the specific activity of the enzyme cannot be estimated since the amount of substrate cleaved appears to vary with the state of folding (Jackson 1983). The maximum processing obtained with detergent-solubilized signal peptidases in the mammalian systems is usually less than 40% (JACKSON and BLOBEL 1980; STRAUSS et al. 1979). Total cleavage can be achieved only when the enzyme is present cotranslationally (LIVELY and WALSH 1983). However, as noted by JACKSON (1983), the enzyme is then restricted to the conditions utilized in the translation reaction. LIVELY and WALSH (1983) have noted that the p-hPL precursor can be more efficiently cleaved if an antibody-precursor complex, formed during translation, is used as the substrate. These data also indicate that the precursor generated in vitro assumes a post-translational conformation that is resistant to in vitro processing by the signal peptidase.

A similar assay has been used to purify the matrix signal peptidase from yeast mitochondria (BOHNI et al. 1983; MCADA and DOUGLAS 1982). The reticulocyte lysate system is used to make the radiolabeled precursor proteins from total yeast mRNA. Since many products are made from the total mRNA, antibody to the appropriate proteins must be added after the addition of the matrix peptidase and before SDS-gel electrophoresis in order to estimate the amount of processing.

The activity of the signal peptidases isolated from bacterial membranes has been mainly estimated from scanning autoradiograms after the separation of the radiolabeled precursor form of the protein from the mature product by SDS-gel electrophoresis. In the case of SPase I, the procoat substrate, synthesized in vitro, is incubated at 37° C for 1 h with a 1% Triton X-100 extract of the enzyme and buffer containing 5 m*M* triethanolamine-Cl, pH 7.5, and 5 m*M* $MgCl_2$. The reaction is terminated with 2X sample buffer, boiled for 3 min, and the products separated by electrophoresis. A unit of activity is defined as the amount of enzyme that produces radioactive bands of procoat and coat on X-ray film of equal intensity (ZWIZINSKI and WICKNER 1980). For SPase II, two different reaction conditions have been employed: (1) radiolabeled prolipoprotein, 0.25% Nikkol, 50 m*M* Tris-HCl buffer (pH 7.4), 0.25% β-mercaptoethanol, and detergent-solubilized enzyme in a final volume of 10 μl (TOKUNAGA et al. 1984) and (2) radiolabeled prolipoprotein, 100 m*M* Tris-acetate (pH 6.0),

0.25% Triton X-100, and detergent-solubilized enzyme (Dev and Ray 1984); the reactions were incubated at 37° C for various periods. The reaction in the former was terminated with 2X SDS sample buffer and in the latter with 80% acetone. The products of the reaction were analyzed by SDS-gel electrophoresis in the phosphate buffer system of Shapiro et al. (1967). Quantitation of enzyme activity was achieved by excising the product bands and counting.

3.2 Signal Peptide Extraction

Another assay for the prolipoprotein signal peptidase (SPase II) was developed during the purification of this enzyme (Dev and Ray 1984). The non-ionic detergent Triton X-100 was used to solubilize the enzyme from the cytoplasmic membrane of *E. coli* and was included in the buffers and assay of this enzyme; however, inclusion of the detergent interfered with the resolution of the precursor and the mature product in the SDS-gel electrophoresis system utilized. In order to remove this detergent before electrophoresis, 80% cold acetone was utilized to terminate the reaction. It was found that even when highly purified preparations of the signal peptidase were used to catalyze the cleavage of prolipoprotein to lipoprotein, the signal peptide was not visible on the autoradiogram. The signal peptide was subsequently localized in the acetone supernatant. It was totally removed from prolipoprotein and lipoprotein which pelleted after centrifugation during the acetone treatment.

In order to test whether the acetone solubility of the signal peptide could be utilized as a quick and convenient assay of SPase II activity, relative to the SDS-gel electrophoresis assay, the two methods were compared. In this assay, the reaction mixture contained in a final volume of 30 µl, [^{35}S]-methionine-labeled modified prolipoprotein, 100 m*M* Tris-acetate (pH 6.0), 0.25% Triton X-100, and purified enzyme (Dev and Ray 1984). The reaction was initiated with substrate and terminated at various times by the addition of 120 µl cold acetone. After the addition of acetone, the tubes were incubated at 4° C for 30 min and then centrifuged. The acetone supernatant was poured quantitatively into a scintillation vial and the radioactivity was determined. The acetone pellet was air dried and suspended in SDS-sample buffer, and products were separated by SDS-gel electrophoresis. After autoradiography, the bands corresponding to the precursor and mature forms of the lipoprotein were excised and the radioactivity was determined. The acetone assay compared well with the gel electrophoresis assay, with a correlation coefficient of 0.86 determined by linear regression. In fact, the deviation from linearity was mainly caused by the data obtained from the gel assay since the standard error for the acetone assay was less than 5%. The validity of this assay was confirmed by showing that globomycin inhibited the cleavage of the signal peptide from the modified prolipoprotein and the unmodified form of prolipoprotein obtained from a gene fusion strain (Tokunaga et al. 1982b) did not release radioactivity into the supernatant.

This assay has been utilized to measure the K_m of SPase II for prolipoprotein and to determine the kinetics of inhibition of SPase II by globomycin (Dev et al. 1985; see Sect. 4.5). We have also utilized the acetone assay to isolate another bacterial signal peptidase that processes the precursor of MBP (RAY et al., manuscript submitted; see Sect. 5).

4 Signal Peptidases

In this section, we will attempt to compare the signal peptidases isolated from various sources as to their purification, properties, and substrate specificities. Even though there is a great deal of variation in the signal sequences of secretory proteins derived from both eukaryotic and prokaryotic sources, certain important features of the cleavage site appear to be maintained and universal since eukaryotic genes coding for a secretory protein can be cloned into *E. coli* (TALMADGE et al. 1980), the protein product secreted and correctly processed; the converse is also true (ROGGENKAMP et al. 1985).

The first signal peptidase to be partially purified was the mammalian enzyme extracted from the microsomes of the dog pancreas (JACKSON and BLOBEL 1977). However, as will be noted, the isolation of signal peptidases from mammalian sources has lagged behind that from the membranes of *E. coli*. The isolation of the mammalian enzyme may be hampered by the post-translational folding of the precursors as noted by JACKSON (1983). It may prove that a bacterial precursor that is stable and can be synthesized in substrate quantities could be utilized for this purpose. It might also be necessary to obtain expression of this enzyme in *E. coli*, if it is possible, to increase the amount of enzyme necessary for purification studies. Two enzymes (SPase I and II) isolated from *E. coli* have been cloned and the corresponding genes have been sequenced (see Sects. 4.3, 4.4).

4.1 Signal Peptidase Activity Isolated from Mammalian Sources

The initial post-translational experiments of JACKSON and BLOBEL (1977) showed that the canine pancreatic signal peptidase could be extracted with detergent from rough microsomes but not from smooth microsomes. This detergent extract (0.5% sodium deoxycholate) correctly processed bovine preprolactin and pre-growth hormone. Subsequent studies (JACKSON and BLOBEL 1980; JACKSON and WHITE 1981; JACKSON 1983) have shown that the signal peptidase from the dog pancreas can be solubilized from microsomal membranes with sodium deoxycholate, Triton X-100, Nikkol, and octyl-glucoside. Concentrations of deoxycholate used to extract the enzyme inhibited the reaction; however, signal peptidase activity could be restored by adding small concentrations of Nikkol (0.04 *M*). STRAUSS et al. (1979, 1980) have also used deoxycholate-solubilized canine pancreatic microsomes as a source of the enzyme to investigate the processing of pre-hPL. Utilizing the post-translational assay, JACKSON and BLOBEL (1980) partially purified the signal peptidase by using DEAE-cellulose column chromatography. In this purification they showed that under low salt conditions,

10 mM triethanolamine-HCl, the signal peptidase was retarded but did not bind to this resin; with the exception of the yeast mitochondrial enzyme, this appears to be true of all signal peptidases. By delipidating the enzyme by molecular seiving in the presence of 0.2% deoxycholate, JACKSON and WHITE (1981) showed that this enzyme required phospholipid for activity and that phosphatidycholine was the most effective in restoring activity. This enzyme was inactivated at 37° C in the presence of deoxycholate and was best assayed at 25° C. Extraction of the enzyme with 1% Triton X-100 is now the method of choice (JACKSON 1983), and the enzyme is assayed in the presence of 2.5 mM soybean phospholipids.

FUJIMOTO et al. (1984) have recently initiated a purification of the signal peptidase from porcine pancreas. Taking advantage of the integral nature of the signal peptidase, they removed many loosely associated membrane proteins from the microsomes by an initial high salt wash followed by a low salt wash containing 0.5% Brij 35. The enzyme was solubilized in 1% Nonidet P-40 and further purified by DEAE-cellulose column chromatography (the enzyme did not bind) and by gel filtration utilizing Sepharose CL-6B. The partially purified enzyme correctly processed pre-hPL to hPL as determined by sequence analysis.

Prepromelittin has been shown to be correctly processed by a signal peptidase obtained by extraction of rat liver rough microsomes with 0.4% deoxycholate after previous washing with high salt and 0.02% deoxycholate (MOLLAY et al. 1982). The enzyme was further purified by centrifugation on a sucrose-deoxycholate density gradient and then reconstituted into phospholipid vesicles. Utilizing the partially purified enzyme, MOLLAY et al. (1982) showed that the precursor could be cleaved and the 21 amino acid signal peptide could be isolated intact using a 1-butanol extraction procedure.

Using a novel purification procedure involving detergent solubilization of hen oviduct microsomes after extraction of the membranes with 0.1 M Na_2CO_3, LIVELY and WALSH (1983) purified a signal peptidase activity that correctly processed pre-hPL. The activity of this enzyme preparation was very low and, as mentioned above, the precursor pre-hPL was a better substrate (60% processing) when complexed to antibody than the uncomplexed pre-hPL (10% processing).

The above mammalian signal peptidase(s) isolated from various sources appear to be similar in many respects (see Table 1). They are integral membrane proteins, with pH optima of 7–8 and are resistant to the inhibition by a number of protease inhibitors (JACKSON and BLOBEL 1980; STRAUSS et al. 1979; FUJIMOTO et al. 1984; LIVELY and WALSH 1983). None of the enzymes have been purified to homogeneity nor have many of their physical and kinetic properties been documented.

4.2 Signal Peptidase Activity Isolated from Yeast Mitochondria

In contrast to the microsomal membrane enzymes, the mitochondrial enzyme is a soluble enzyme located in the mitochondrial matrix. This enzyme has been purified from yeast approximately 100-fold by BOHNI et al. (1983) and 200-fold

Table 1. Comparison of properties of various signal peptidases

	SPase I (1, 10, 12, 13, 14, 15, 16)	SPase II (2, 17)	SPase III (3, unpublished)	Porcine pancreas (4)	Canine pancreas (5, 6, 11)	Hen oviduct (7)	Yeast mitochondria (8, 9)
1. Purity	6000 ×	35000 ×	~20000 ×	Partially	Partially	Purified	Purified-200 ×
2. Molecular weight	37000	17000	60000	ND	55A in detergent	ND	115000
3. Substrates: (–) indicates nonsubstrate	p-Coat protein p-MBP, p-IgCk Others reported synthetic peptides (–)	Prolipoprotein p-MBP (–)	p-MBP prolipoprotein (–)	p-hPL Azocasein (–)	p-hPL, p-GH p-Prolactin aminocoumarin peptides (+ –)	p-hPL	p-Matrix p-Inter membrane Nonmitochondrial p-proteins (–)
4. % Substrate Cleaved post-translationally	100	100	100	30	15–40	10–20	100
5. Inhibitors	1, 10-Ptn DNP, NEM high NaCl Mg^{2+}	Globomycin PEA, NEM, PMSF, TAME Hg^{2+} Pepstatin Chymostatin	1, 10-Ptn EDTA, Hg^{2+} 1% DOC Leupeptin Chymostatin	Chymostatin p-Ptn	Chymostatin (200–600 μg/ml) 1% DOC	1% DOC	EDTA O-Ptn, 1, 10-Ptn GTP
6. Noninhibitors	Serine esterase EDTA, o-Ptn Peptides	1, 10-Ptn, EDTA Peptides Trypsin inhib.	PMSF Pepstatin Antipain TLCK	Serine esterase Peptides Globomycin	Serine esterase Peptides, EDTA Trypsin inh., NEM	ND	Serine esterase Peptides
7. Activators	ND	1 m*M* DTT	$Ca^{2+} > Mg^{2+} > Mn^{2+}$	ND	Phospholipid (PC)	PC	Co^{2+}, Mn^{2+}, Zn^{2+}
8. pH optimum	8.5–9	6	6–8	7–8	ND	ND	7–8
9. K_m	ND	6 μ*M*	0.5–1.0 μ*M*	ND	ND	ND	ND

DNP, dinitrophenol; PEA, phenethyl alcohol; NEM, ethylmaleimide; o-PTN, *o*-phenanthroline; PMSF, phenylmethylsufonylfluride; TAME, tosyl arginyl methyl ester; TLCK, p-tosyl-L-lysine chloromethyl ketone; DTT, dithiothreitol; ND, not determined; p, precursor

References: (1) Zwizinski and Wickner (1980); (2) Dev and Ray (1984); (3) (Ray et al., (to be published); (4) Fujimoto et al. (1984); (5) Jackson and Blobel (1980); (6) Strauss et al. (1979); (7) Lively and Walsh (1983); (8) McAda and Douglas (1982); (9) Bohni et al. (1983); (10) Wolfe et al. (1983a); (11) Jackson and White (1981); (12) Ito (1982); (13) Watts et al. (1983); (14) Wolfe et al. (1982); (15) Zwinzinski et al. (1981); (16) Wolfe et al. (1983b); (17) (Dev et al. (1985)

by McAda and Douglas (1982). This enzyme cleaves mitochondrial precursors, initially synthesized in the cytoplasm, destined for the mitochondrial matrix, as well as the inner membrane and the intermembrane space. In agreement with these proposed functions, the purified enzyme cleaves pre-matrix proteins (i.e., precursors of the α and β subunits of the F_1-ATPase, cytochrome oxidase, and citrate synthetase) to their mature forms and cleaves the precursors of outwardly facing inner membrane proteins (i.e., cytochrome c_1) and precursors of soluble proteins in the intermembrane space (i.e., cytochrome b_2) to an intermediate form which is later processed further. It is proposed that a second protease located in the inner membrane cleaves the intermediate form to the mature form. This second activity has yet to be identified. The purified matrix SPase has a pH optimum of 7.5. Due to its inhibition by 1,10-phenanthroline, EDTA and guanosine 5-triphosphate (GTP), it is thought to be a metalloenzyme. The effect of these inhibitors can be reversed by Zn^{2+}, Co^{2+}, or Mn^{2+}. Like the mammalian enzymes, this enzyme is not sensitive to serine protease inhibitors, peptide inhibitors, or trypsin inhibitors. From the work of McAda and Douglas (1982), the soluble enzyme elutes from a molecular sieving column with an apparent molecular weight of 115000 whose subunit molecular weight determined in SDS-gels may be 59000. The properties of these enzymes are also shown in Table 1. It is interesting to note that the enzyme exhibits 100% processing of the mitochondrial precursors tested; this finding is not surprising since it is known that the translocation and processing of the mitochondrial precursors imported from the cytoplasm is a post-translational event.

4.3 Signal Peptidase I Isolated from *E. coli*

The first bacterial signal peptidase was isolated and purified by Zwiziski and Wickner (1980) from the membranes of *E. coli*. This initial purification employed extraction with 1% Triton X-100, followed by DEAE-cellulose column chromatography, selective ethanol precipitation, and chromatography on butyl agarose. The enzyme was identified as a 37000 molecular weight protein after electrophoresis on Triton gels and further analysis of the active fractions by SDS-gel electrophoresis. The assay of this unique enzyme involves the post-translational cleavage of the M13 procoat protein to coat protein as assayed by SDS-gel electrophoresis. The purified enzyme processed the precursor form to the correct mature form as evidenced by sequence analysis of the released signal peptide and the mature protein. This enzyme was localized equally between the inner and outer membrane (Zwizinski et al. 1982) and Date (1983) has shown that SPase I is an essential enzyme.

Wickner and coworkers (Date and Wickner 1981; Zwizinski et al. 1981; Wolfe et al. 1982, 1983a; Silver and Wickner 1983) have partially characterized the enzyme, cloned the gene for SPase I or leader peptidase, mapped the gene on the *E. coli* chromosome at 55 min, and obtained the nucleotide sequence for the gene that codes for SPase I. The gene codes for a protein of 323 amino acids with a molecular weight of 35994, which is almost identical to the molecular weight determined by SDS-gel electrophoresis. The enzyme is not synthesized as a precursor in that it does not have a cleavable leader sequence; however, it may follow the same route of export as precursor proteins since the insertion of SPase I into the membrane is blocked in *secA* and *secY* mutants of *E. coli* (WOLFE et al. 1985). Most of the polypeptide is exposed on the outer surface of the inner membrane and is anchored by a membrane-spanning segment located near the amino-terminal region (WOLFE et al. 1983a). The enzyme has been reported to process the precursor form of a wide variety of periplasmic proteins and outer membrane proteins, including the precursor forms of M13 procoat, leucine-specific and isoleucine/valine binding proteins, OmpA, protein III, LamB, and MalE (WOLFE et al. 1982), as well as the precursor of the C_k fragment of human immunoglobulin (WATTS et al. 1983).

This enzyme resembles the mammalian signal peptidases in that it is not inhibited by a wide variety of protease inhibitors (see Table 1) nor does it bind to DEAE-cellulose. The enzyme has a pH optimum between 8.5 and 9.0, is inhibited by NaCl above 160 m*M*, and is inhibited by Mg^{2+} above 1 m*M*; however, the activity is inhibited by 1,10-phenanthroline but not by EDTA, *o*-phenanthroline, or 2,6-pyridine dicarboxylic acid. The purified enzyme has been utilized by ITO (1982) to process pre-MBP and by WATTS et al. (1983) to process the precursor of the C_k fragment of human immunoglobulin. However, it is important to note that the rates at which these substrates are processed in vitro are very slow. ITO (1982) purified the precursor of MBP by affinity chromatography (FERENCI and KLOTZ 1978) and processed this substrate (900 ng = 0.021 nmol) with 1.7–170 ng (0.00004–0.004 nmol) of purified enzyme for 1 h at 37° C. The extent of the conversion of pre-MBP to MBP was dependent on the amount of enzyme used; however, even after 1 h at the highest enzyme concentration, approximately 20% of the substrate was not utilized. In our hands, SPase I does process pre-MBP but at a very slow rate (see Sect. 5). Likewise, the amount of purified SPase I (1.3 and 33 μg) utilized to process the C_k fragment precursor that was prepared in an in vitro reaction (where the specific activity of the substrate should be very high) indicated that the processing reaction was enzyme concentration dependent but was very slow. This slow processing might be accounted for by the post-translational folding of the substrate as previously discussed (Sect. 2.1).

4.4 Signal Peptidase II Isolated from *E. coli*

The murein lipoprotein or Braun's lipoprotein contains a unique modified cysteine residue at the amino terminus of the mature protein. This specific lipoprotein and other lipoproteins in *E. coli* (ICHIHARA et al. 1981) are synthesized as precursor proteins (INOUYE et al. 1977); the biogenesis of the mature lipoprotein involves modification before the processing reaction is initiated. TOKUNAGA et al. (1982a) showed that prolipoprotein signal peptidase (SPase II) was distinct from SPase I (see above). The structural gene for SPase II (designated *lsp*) has been cloned by two independent groups (TOKUNAGA et al. 1983; YAMAGATA et al. 1983a), genetically mapped (REGUE et al. 1984; YAMAGATA et al. 1983b), and its nucleotide sequence determined (INNIS et al. 1984; YU et al. 1983). WU and coworkers (Innis et al. 1984; Tokunaga et al. 1985; Kamio et al. 1985) have since shown that the *lsp* gene is transcribed on the same mRNA as *ileS* (isoleucyl-tRNA synthetase) and is under the promotional control of gene X that codes for a cytoplasmic protein with an apparent molecular weight of 35000. From the amino acid sequence deduced from the DNA sequence, the *lsp* gene encodes for a protein containing 164 amino acids and having a molecular weight of 18000. A model suggested by a computer-assisted analysis of the secondary structure suggests that there are four hydrophobic domains that share features common to transmembrane segments of a number of integral membrane proteins, indicating that the protein spans the membrane four times. Two types of evidence (Sect. 1) suggest that the *lsp* gene product is an essential enzyme.

Dev and Ray (1984) have purified and partially characterized this unique signal peptidase from the membranes of *E. coli* B. The purified SPase II correctly processes the prolipoprotein to its mature form plus an intact signal peptide. In confirmation of earlier observations of Tokunaga et al. (1982b), they showed that the modified prolipoprotein is the natural substrate for this enzyme and that the unmodified precursor of lipoprotein was not a substrate. Furthermore, pre-maltose binding protein is not processed by this enzyme nor is there any nonspecific proteolytic activity associated with it. The monomeric molecular weight of this enzyme, as determined by SDS-gel electrophoresis, is 17800, which is not significantly different than the molecular weight deduced from the nucleotide sequence (Innis et al. 1984). Moreover, the in vitro product of a coupled transcription-translation system programmed by plasmid DNA carrying the *lsp* gene was shown to comigrate with the above-purified wild-type enzyme (Tokunaga et al. 1985).

SPase II is localized in the inner membrane (Tokunaga et al. 1984; Dev and Ray 1984). The purified enzyme exhibits no dependence on divalent cations for enzymatic activity. The purified enzyme is stable at 4° C or at −20° C when stored in buffers containing 10% glycerol, 1% Triton X-100 and 1 m*M* dithiothreitol (DTT). The enzyme is not inhibited by a number of protease inhibitors (see Table 1) but is very sensitive to inhibition by globomycin. Globomycin does not inhibit SPase II by binding to the substrate. By utilizing the purified SPase II and the prolipoprotein substrate with known specific radioactivity, Dev et al. (1985) have shown that the K_m of SPase II for the substrate is 6 μ*M*. Furthermore, these authors have shown that globomycin inhibits SPase II by binding to the enzyme in a noncompetitive manner with a K_i of 36 n*M* and that globomycin does not bind to the substrate. Antibody to the cloned enzyme is not yet available.

5 Evidence for Another Signal Peptidase Isolated from *E. coli*

As shown above there are two signal peptidases thus far isolated from the membranes of *E. coli* and both have been shown to be essential enzymes. However, it remains to be established if these are the only signal peptidases responsible for processing or if there are other signal peptidases also present in *E. coli*. In order to isolate signal peptidases, there must be a good source of radiolabeled precursors available as substrates.

The isolation and characterization of a number of signal sequence mutations in the *malE* gene, encoding MBP, now provide a good source of radiolabeled MBP precursors (BEDOUELLE et al. 1980; BANKAITIS et al. 1985; see Sect. 2.4). In most cases, a mutation or an in-frame deletion within the coding region of the signal sequence results in the synthesis of an export-deficient MBP precursor that accumulates in the cytoplasm (EMR and BASSFORD 1982). Since the isolation of signal sequence mutations in the *malE* gene (see Fig. 1), BASSFORD and coworkers (BASSFORD and BECKWITH 1979; EMR and BASSFORD 1982; BASSFORD et al. 1984; BANKAITIS et al. 1984; BANKAITIS and BASSFORD 1985) have shown that extragenic *prl* mutations could phenotypically suppress certain signal sequence mutations and allow the correct localization and processing of MBP. These findings indicated that the various precursors synthesized in the signal sequence mutants could be utilized as substrates for the isolation of signal peptidases.

As mentioned previously, MBP signal sequence mutants provide precursors synthesized in vivo with a specific activity of $1–3 \times 10^7$ cpm/nmol. Utilizing these radiolabeled substrates, particularly the substrate derived from the *malEΔ12–18* mutant (see Fig. 1), another signal peptidase has been isolated from the membranes of *E. coli* B that can process the MBP precursors (RAY et al., to be published).

While purifying SPase II, initial Triton X-100 extracts of this enzyme preparation were also tested for their ability to process pre-MBP to its mature product, as analyzed by SDS-gel electrophoresis. It was found that an enzyme activity was present that processed the mutant precursor Δ12–18; this activity initially was assumed to represent SPase I. This activity was further purified by acid precipitation, DEAE-column chromatography, Sepharose 4BCl column chromatography, and Chromatofocusing utilizing the Δ12–18 precursor as the substrate. As shown in Fig. 2A, this enzyme preparation processes each of the MBP precursors shown in Fig. 1 to the apparent correct mature product as determined by their mobility after electrophoresis on SDS gels. In contrast, as shown in Fig. 2B, purified SPase I (obtained from W. Wickner) did not process or processed very slowly the MBP precursors Δ12–18, 18-1, and 19-1. This new enzyme activity has been designated SPase III and correctly processes the wild-type precursor and the 16-1 precursor as determined by amino-terminal sequence analysis of the matured product (RAY et al., to be published). The wild-type substrate synthesized in vitro is also processed by SPase III (C. MacGregor, unpublished results).

The wild-type MBP precursor was found to be processed by both SPase I and SPase III. However, as shown in Fig. 3, SPase III processes the wild-type precursor much faster than does purified SPase I. It was estimated that the SPase III enzyme preparation contained only 0.008 μg protein per time point, while 4 μg SPase I was utilized per time point. The above data show that, with the wild-type precursor of MBP as the substrate, SPase III is 500-fold more active than SPase I. To illustrate further that the enzymes are different

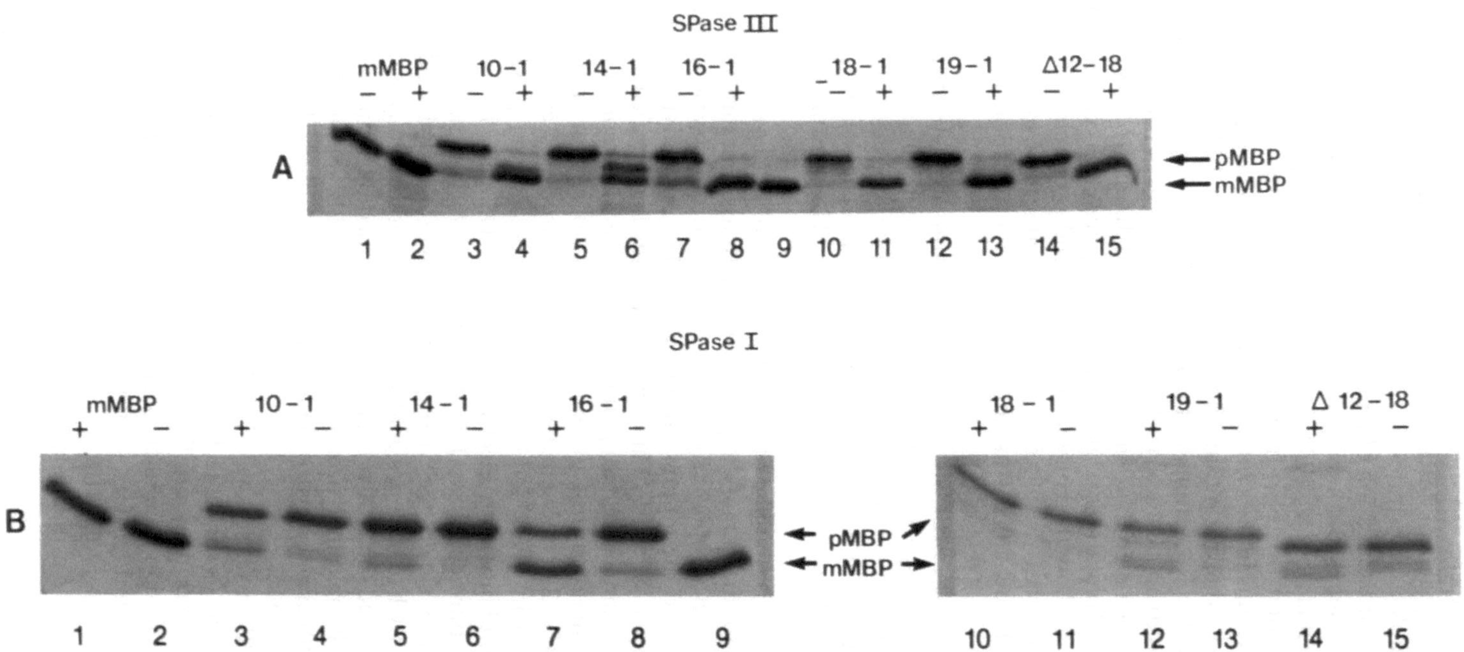

Fig. 2A, B. Processing of the precursors isolated from the signal sequence mutants by SPase III and SPase I. The various precursors were processed in a reaction volume of 45 µl containing 15 µl buffer, 15 µl substrate (about 50000 CPM), and 15 µl enzyme (5 µg SPase I; 0.01 µg SPase III) or buffer in the controls for 1 h at 37° C. The reaction was terminated as described. **A** Processing with SPase III. Lanes 3, 5, 7, 10, 12, and 14 are controls without enzyme; lanes 4, 6, 8, 11, 13, and 15 are aliquots of the processed precursors containing substrate obtained from mutants 10-1, 14-1, 16-1, 18-1, 19-1, and Δ12–18, respectively. Lanes 1 and 9 are mMBP without added enzyme and lane 2 is mMBP with added enzyme. **B** Processing with SPase I. Lanes 4, 6, 8, 11, 13, and 15 are controls without added enzyme. Lanes 3, 5, 7, 10, 12, and 14 are aliquots of the processed precursors containing substrate obtained from mutants 10-1, 14-1, 16-1, 18-1, 19-1, and Δ12–18 respectively. Lanes 2 and 9 are mMBP without added enzyme and lane 1 is mMBP with added enzyme

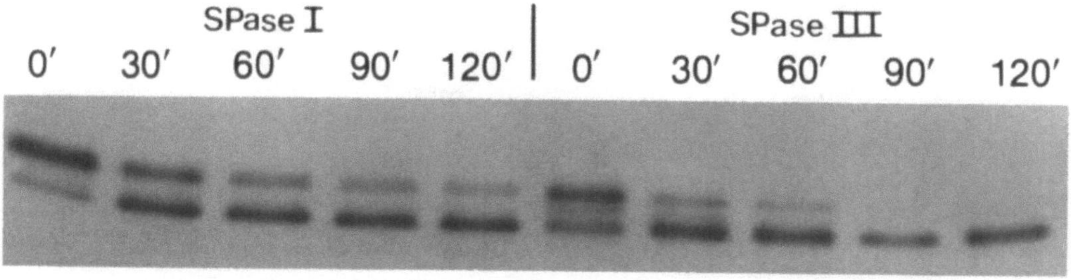

Fig. 3. Time course hydrolysis of the normal precursor of MBP by SPase 1 and SPase III. The normal precursor was processed by both enzymes in a final reaction volume of 180 μl containing 60 μl buffer, 60 μl substrate (300000 cpm), and 60 μl enzyme. The reaction was initiated by the addition of enzyme (approximately 24 μg SPase I and 0.048 μg SPase III) at 37° C. Aliquots of 30 μl were withdrawn at the times indicated and diluted with 120 μl cold acetone and analyzed as described. Lanes 1–5 correspond to the processing by SPase I and lanes 6–10 correspond to processing by SPase III

and that the activities are just not a reflection of the isolation procedure, samples of each enzyme utilized in the time course assay (Fig. 3) were analyzed by SDS-gel electrophoresis and the protein profiles were compared. As can be seen in Fig. 4, the protein profiles are very different (lanes 2 and 4 contain SPase I, lanes 3 and 5 contain SPase III). The lanes containing SPase III do not contain any silver-stained protein at the molecular weight of 36000, which is the known molecular weight of SPase I.

SPase III has been purified to near homogeneity (P.H. Ray, C. Jones, I.K. Dev, J. Fikes, V. Bankaitis, and P.J. Bassford, Jr., manuscript in preparation) and appears to be a protein with a subunit molecular weight of 58000–60000. The enzyme does not process either modified or unmodified prolipoprotein. The K_m of SPase III for the MBP 16-1 precursor is 0.5–1.0 μ*M*. This signal peptidase is inhibited by EDTA and 1,10-phenanthroline and requires Ca^{2+}, Mg^{2+}, or Mn^{2+} (Ca^{2+} preferred) for activity. Like other signal peptidases, it does not appear to be inhibited by the normal protease inhibitors shown in Table 1 and is purified in the presence of 0.2 m*M* phenylmethylsulfonylfluoride (PMSF). Antibodies are being raised to the purified protein. We are currently investigating the processing of other *E. coli* protein precursors by SPase III. The role of SPase III in protein export in *E. coli* remains to be established.

6 Signal Sequences and Processing Sites

The role of the signal sequence in protein export has been well documented (see Silhavy et al. 1983; Bankaitis et al. 1985; Benson et al. 1985). There is little primary sequence homology between the 50–60 (Watson 1984) known eukaryotic and prokaryotic signal sequences. As previously mentioned, the protein products of non-lipoprotein prokaryotic genes expressed in eukaryotes are correctly processed (Roggenkamp et al. 1985) and the converse is also true

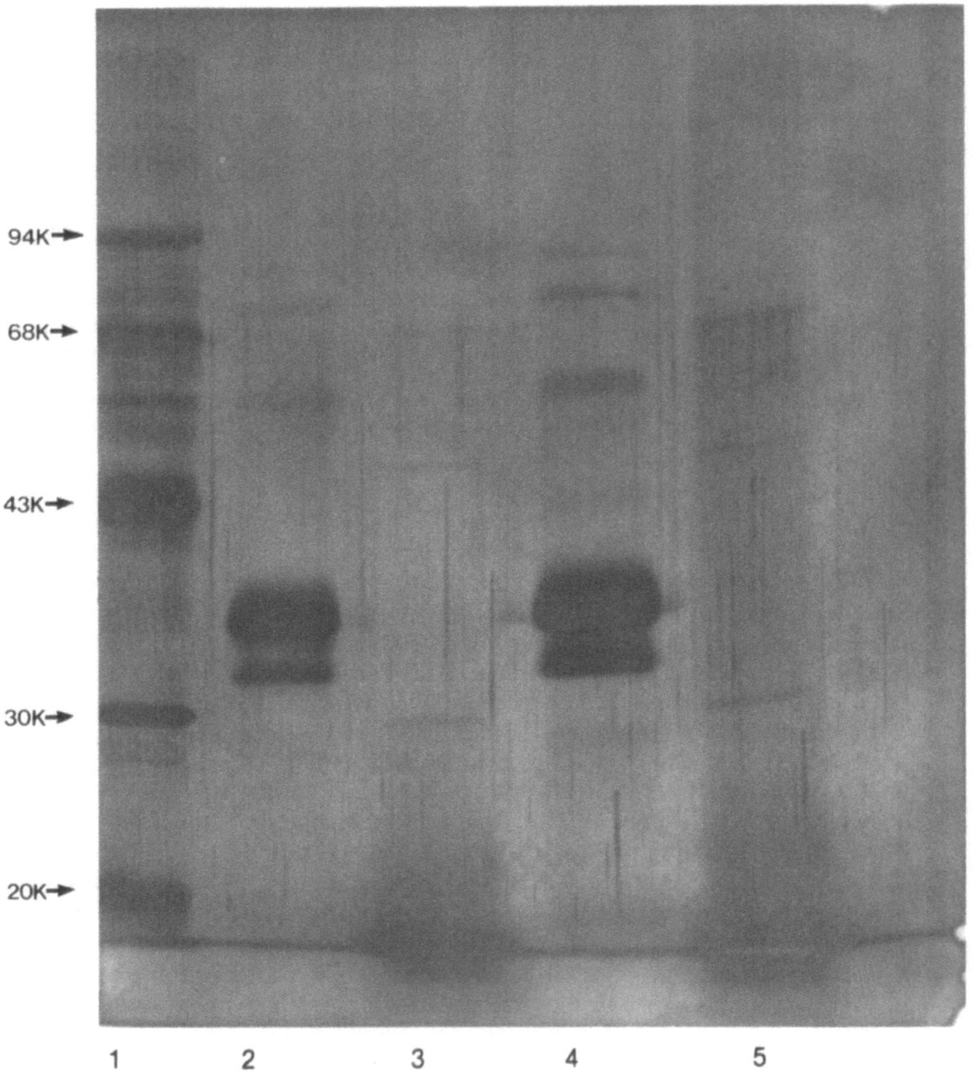

Fig. 4. Protein profile of SPase III compared with SPase I. Twenty-five and 50 μl of SPase III (five to ten times the amount of enzyme utilized in the processing reactions) representing approximately 0.04 and 0.08 μg protein was analyzed by SDS-polyacrylamide gel electrophoresis using 10% gels, lanes 3 and 5 respectively. Lane 1 contains the low molecular weight standards from Pharmacia, lanes 2 and 4 contain 5 and 10 μg SPase I. The protein profile was detected using the Silver Stain Kit supplied by Bio-Rad

(TALMADGE et al. 1980). If there is little primary sequence homology between signal peptides, how then are the secretory proteins that contain cleavable signal peptides correctly recognized and processed? Both PERLMAN and HALVORSON (1983) and VON HEIJNE (1984) have recently addressed this question. As noted by others (SILHAVY et al. 1983; BANKAITIS et al. 1985; PERLMAN and HALVORSON

1983), signal peptides have three properties in common: (1) at the amino terminus one to five positively charged residues, either lysine or arginine, are found; (2) the last positively charged residue is followed by a hydrophobic core of 8–13 hydrophobic amino acids arranged in a nonrandom sequence and probably in an α-helical configuration; and (3) the processing site is immediately preceded by a small uncharged amino acid, usually alanine or glycine. According to the model of PERLMAN and HALVORSON (1983), the nonlipoprotein signal peptidases are targeted to the cleavage site by a consensus cleavage site A(-3)-X(-2)-B(-1) immediately preceding the first amino acid of the mature protein. A and B are small, neutral amino acids usually alanine, serine, or glycine. The most common consensus sequence is alanine-X-alanine. From the data analyzed, PERLMAN and HALVORSON (1983) also noted that the region -6 to -3 in the signal sequence always showed a tendency to form a β-turn, which they suggested was necessary to bring the cleavage site into juxtaposition for cleavage by the signal peptidase. Their model depicts the amino-terminus containing the positively charged residues binding to some component on the inner surface of the cytoplasmic membrane and the hydrophobic core, in an α-helical conformation, extending through the bilayer. The region at -6, from the first amino acid of the mature protein would elicit a β-turn region and position the cleavage site on the external side of the membrane for correct processing. This model appears to account for most data derived from the analysis of the signal sequences.

Are the requirements for the cleavage of the nonlipoprotein secretory proteins the same as for the cleavage of prolipoproteins by SPase II? The answer is probably no. The cleavage sites of the known prolipoproteins (see PUGSLEY and SCHWARTZ 1985) are similar to each other but not identical in that the -3 to $+1$ regions contain the following sequences: leu-ala-gly-cys; leu-ser-ala-cys; leu-ser-gly-cys; leu-val-gly-cys; or leu-val-ala-cys. The important component of these sequences is the cysteine residue, which must be fully modified (TOKUNAGA et al. 1982b, 1984) before the precursor of Braun's prolipoprotein can be processed (TOKUNAGA et al. 1982; DEV and RAY 1984). The fact that the unmodified precursor is not processed even though the amino acids in the signal sequence and the remainder of the mature protein are the same indicates that the modification reactions and the resulting conformational changes or hydrophobicity changes resulting from these modifications are most important for the cleavage to occur. INOUYE and coworkers have used site-specific mutagenesis to modify the prolipoprotein signal sequence (INOUYE et al. 1982, 1984; VLASUK et al. 1983, 1984). The results have shown that this signal sequence can tolerate numerous changes; however, any alteration that prevents the modification of the precursor also results in a precursor that cannot be processed by SPase II.

7 Fate of the Signal Peptide

In *E. coli*, genetic studies have shown that the signal peptide is essential for the initiation of export across the cytoplasmic membrane (SILHAVY et al. 1983;

Benson et al. 1985). However, the fate of the signal peptide after its cleavage from the precursors is not known. Is the signal peptide released into the periplasm, degraded in the membrane, or secreted back into the cytoplasm and degraded there? It is known that the signal peptide is degraded very quickly (Jackson and Blobel 1977; Habener et al. 1979; Zwizinski and Wickner 1980; Hussain et al. 1982b). The reason for its rapid degradation has been elucidated by the recent work of Austen et al. (1984) and that of Koren et al. (1983). Their data suggest that the signal peptide itself may be involved in the regulation of the synthesis of secretory proteins and in the secretory process itself. If this view is correct, then the rapid removal and degradation of the cleaved signal peptide is necessary to maintain proper export.

7.1 Localization of Signal Peptide Hydrolases

Wickner and coworkers (Zwizinski and Wickner 1980; Silver et al. 1982) initially described a signal peptide hydrolase activity localized in the soluble fraction (S-100) of *E. coli* extracts which was responsible for the degradation of the signal peptide generated during the processing of the M13 procoat protein. A membrane-bound signal peptide peptidase (or hydrolase) that degrades the signal peptide derived from the correct processing of prolipoprotein, generated in situ, has also been observered in *E. coli* (Hussain et al. 1982b). This hydrolase is inhibited by a number of protease inhibitors. When prolipoprotein-containing membranes, generated in the presence of globomycin, were incubated in the presence of antipain, leupeptin, chymostatin, and elastatinal, a small peptide appeared on polyacrylamide gels at the position where the signal peptide should migrate if it were not degraded.

Using the signal peptide isolated from the purified prolipoprotein and the purified SPase II, Novak et al. (Novak et al., 1985) have shown that the signal peptide is degraded (at least 400-fold) more rapidly than other cellular proteins in *E. coli.* Greater than 90% of this signal peptide degrading activity is localized in the cytoplasm and only 10% of this activity is found in the membrane fraction. Since most of the signal peptide hydrolase activity is localized in the cytoplasm, it is tempting to speculate that the role of the positively charged amino-terminus of the signal peptide is to direct the signal peptide back into the cytoplasm, whereby it can be fully degraded.

7.2 Specific Signal Peptide Hydrolases

The signal peptide hydrolyzing activity in the cytoplasm has been purified by Novak et al. (1985) and shown to be effected by two separate proteases. One of these signal peptide hydrolases is identical to protease So, which has been characterized by Chung and Goldberg (1983), and this protein represents less than 10% of the cytoplasmic activity. The majority of the cytoplasmic signal peptide hydrolase activity is associated with a 68000 molecular weight protein which is similar to the oligopeptidase described by VIMR et al. (1982). This similarity is based on the fact that the enzyme degrades *N*-acetyl-$(Ala)_4$ but does not hydrolyze *N*-acetyl-$(Ala)_3$, and that the activity is inhibited by EDTA and stimulated by Co^{2+}. Also, a mutant of *S. typhimurium* that is deficient in the oligopeptidase activity is deficient in the 68000 molecular weight cytoplasmic signal peptidase activity (DEV et al. to be published).

Recently, ICHIHARA et al. (1984) have provided convincing evidence that a membrane-bound protease initially purified by PACAUD (1982) is the enzyme responsible for the disappearance of the signal peptide generated in situ by the processing of prolipoprotein. This enzyme was designated protease IV by PACAUD (1982). Both the signal peptide hydrolase activity and protease IV activities copurified and were inhibited to similar degrees by antipain, leupeptin, and chymostatin.

7.3 Degradation of the Signal Peptide

The signal peptide of prolipoprotein is a 20 amino acid peptide that has only one methionine residue and that is at the amino terminus. The products generated at various times during the hydrolysis of the $[^{35}S]$-methionine-labeled signal peptide by cytoplasmic as well as membrane extracts were analyzed by both gel filtration and C-18 μBondapak reverse-phase high-pressure liquid chromatography (NOVAK et al. 1985; DEV et al. to be published). The results showed that initially a number of $[^{35}S]$-methionine-labeled peptides of different molecular weights and different degrees of hydrophobicity were generated during the hydrolysis reaction, which eventually yielded a peak of radioactivity corresponding to $[^{35}S]$-methionine in a time-dependent manner. If the signal peptide were degraded from the amino terminus, then only one labeled small molecular weight product would have been detected. If the peptide were degraded from both the ends, then methionine should be released very quickly and this was not the case. Also, if the signal peptide were degraded from the amino terminus end, the prolipoprotein precursor itself should be unstable and degraded in a time-dependent manner and this also was not the case. These data suggest that the signal peptide is degraded from the carboxyl-terminus, in a stepwise manner. It is not clear whether the stepwise degradation of the signal peptide releases one, two, or possibly three amino acids at a time. It is possible that the signal peptide is clipped into two fragments, then four, etc. and then the smaller fragments are degraded, simply depending upon the affinity of the enzyme for the various fragments.

It appears that there are three possibilities regarding the degradation of the signal peptide: (1) the in vivo activity exhibited by protease IV is sufficient to degrade all the signal peptide generated during protein export; (2) that protease IV initially cleaves the signal peptides and the products of the initial cleavage can then diffuse back into the cytoplasm and be degraded by So and the oligopeptidase; or (3) that the signal peptides are degraded by any one of the three proteins that exhibit signal peptide hydrolase activity.

8 Summary

At this time it is difficult to compare the signal peptidases isolated from the various sources as to their physical properties, substrate specificities, and inhibitor profiles, simply because the data are not available. All the signal peptidases appear to be tightly membrane bound, with the exception of SPase I and the yeast mitochondrial enzyme; all appear not to bind to DEAE-cellulose with the exception of the yeast mitochondrial enzyme; all appear to be resistant to most normal protease inhibitors; and all appear to be metallo-endopeptidases, with the exception of SPase II. The mammalian SPases differ from the bacterial SPases in their requirement for phospholipid and their more stringent requirement for the appropriate substrate conformation.

In the bacterial system, it is clear that the signal peptidases are essential enzymes and are required for normal localization and activity. In mammalian systems, there are no data to support this view; however, it can be imagined that in the absence of processing the precursors may not be properly released into the lumen of the rough endoplasmic reticulum and that further processing and glycosylation would be impaired. In bacteria, the recent identification of SPase II, which functions only in the cleavage of the prolipoprotein class of proteins, SPase I, which appears to be a general nonlipoprotein signal peptidase, and SPase III, which processes at least one precursor whose mature product is localized in the periplasm, suggests that there may be several SPases that function in the processing of secretory proteins. Clearly, this is the case in the yeast mitochondria. Whether this situation exists in the eukaryotic microsomes cannot be determined until pure SPases have been isolated and their kinetic properties studied using a number of precursors with known specific activities.

References

Austen BM, Hermon-Taylor J, Kaderbhai MA, Ridd DH (1984) Design and synthesis of a consensus signal sequence that inhibits protein translocation into rough microsomal vesicles. Biochem J 224:317–325

Bankaitis VA, Bassford PJ Jr (1985) Proper interaction between at least two components is required for efficient export of proteins to the *Escherichia coli* cell envelope. J Bacteriol 161:169–178

Bankaitis VA, Rasmussen BA, Bassford PJ Jr (1984) Intragenic suppressor mutations that restore export of maltose binding protein with a truncated signal peptide. Cell 37:243–252

Bankaitis VA, Ryan JP, Rasmussen BA, Bassford PJ Jr (1985) The use of genetic techniques to analyze protein export in *Escherichia coli*. In: Knauf PA, Cook JS (eds) Current topics in membranes and transport, vol 24. Academic Press Orlando, Florida, pp 105–150

Bassford P, Beckwith J (1979) *Escherichia coli* mutants accumulating the precursor of a secreted protein in the cytoplasm. Nature 277:538–541

Bassford PJ Jr, Silhavy TJ, Beckwith JR (1979) Use of gene fusion to study secretion of maltose-binding protein into *Escherichia coli* periplasm. J Bacteriol 139:19–31

Bassford PJ Jr, Bankaitis VA, Rasmussen BA, Ryan JP (1984) Genetic studies on the mechanism of secretion of maltose-binding protein of *E. coli*. J Cell Biochem 24:303–309

Bedouelle H, Bassford PJ Jr, Fowler AV, Zabin I, Beckwith J, Hofnung M (1980) Mutations which alter the function of the signal sequence of the maltose binding protein of *Escherichia coli*. Nature 285:78–81

Benson SA, Hall MN, Silhavy TJ (1985) Genetic analysis of protein export in *Escherichia coli* K-12. Ann Rev Biochem 54:101–134

Blobel G, Dobberstein B (1975) Transfer of proteins across membranes. J Cell Biol 67:835–851

Bohni PC, Daum G, Schatz G (1983) Import of proteins into mitochondria. J Biol Chem 258:4937–4943

Casadaban MJ (1976) Transposition and fusion of the lac operon to selected promoters in *E. coli* using bacteriophage lambda and Mu. J Mol Biol 104:541–555

Chen L, Rhoads D, Tai PC (1985) Alkaline phosphatase and OmpA protein can be translocated posttranslationally into membrane vesicles of *Escherichia coli*. J Bacteriol 161:973–980

Chung HC, Goldberg AL (1983) Purification and characterization of protease So, a cytoplasmic serine protease in *Escherichia coli*. J Bacteriol 154:231–238

Dalbey RE, Wickner W (1985) Leader peptidase is required for the release of exported proteins from the outer surface of the *Escherichia coli* plasma membrane. J Biol Chem 260:15925–15931

Daniels CJ, Bole DG, Quay SC, Oxender DL (1981) Role of membrane potential in the secretion of protein into the periplasm of *Escherichia coli*. Proc Natl Acad Sci USA 78:5396–5400

Date T (1983) Demonstration by a novel genetic technique that leader peptidase is an essential enzyme of *Escherichia coli*. J Bacteriol 154:76–83

Date T, Wickner W (1981) Isolation of the *Escherichia coli* leader peptidase gene and effects of leader peptidase overproduction in vivo. Proc Natl Acad Sci USA 78:6106–6110

Dev IK, Ray PH (1984) Rapid assay and purification of a unique signal peptidase that processes the prolipoprotein from *Escherichia coli* B. J Biol Chem 259:11114–11120

Dev IK, Harvey RJ, Ray PH (1985) Inhibition of prolipoprotein signal peptidase by globomycin. J Biol Chem 260:5891–5894

Dev IK, Harvey RJ, Ray PH (1986) Identification of the oligopeptidase as the major cytoplasmic signal peptide hydrolase in *Salmonella typhimurium*. Unpublished data

Emr SD, Bassford PJ (1982) Localization and processing of outer membrane and periplasmic proteins in *Escherichia coli* strains harboring export-specific suppressor mutations. J Biol Chem 257:5852–5860

Emr SD, Silhavy TJ (1980) Mutations affecting localization of an *Escherichia coli* outer membrane protein, the bacteriophage lambda receptor. J Mol Biol 141:63–90

Emr SD, Silhavy TJ (1982) Molecular components of the signal sequence that function in the initiation of protein export. J Cell Biol 95:689–696

Ferenci T, Klotz U (1978) Affinity chromatographic isolation of the periplasmic maltose binding protein of *Escherichia coli*. FEBS 94:213–217

Ferenci T, Randall LL (1979) Precursor maltose-binding protein is active in binding substrate. J Biol Chem 254:9979–9981

Fujimoto Y, Watanabe Y, Uchida M, Ozaki M (1984) Mammalian signal peptidase: partial purification and general characterization of the signal peptidase from microsomal membranes of porcine pancreas. J Biochem 96:1125–1131

Goodman JM, Watts C, Wickner W (1981) Membrane assembly: posttranslational insertion of M13 procoat protein into *E. coli* membranes and its proteolytic conversion to coat protein in vitro. Cell 24:437–441

Habener JF, Rosenblatt M, Dee PC, Potts JT Jr (1979) Cellular processing of pre-proparathyroid hormone involves rapid hydrolysis of the leader sequence. J Biol Chem 254:10596–10599

Halegoua S, Inouye M (1979) Translocation and assembly of outer membrane proteins of *Escherichia coli*. Selective accumulation of precursors and novel assembly of intermediates caused by phenethyl alcohol. J Mol Biol 130:39–61

Hussain M, Ichihara S, Mizushima S (1980) Accumulation of glyceride-containing precursor of the outer membrane lipoprotein in the cytoplasmic membrane of *Escherichia coli* treated with globomycin. J Biol Chem 255:3707–3712

Hussain M, Ichihara S, Mizushima S (1982a) Mechanism of signal peptide cleavage in the biosynthesis of the major lipoprotein of the *Escherichia coli* outer membrane. J Biol Chem 257:5177–5182

Hussain M, Ozawa Y, Ichihara S, Mizushima S (1982b) Signal peptide digestion in *Escherichia coli*. Eur J Biochem 129:233–239

Ichihara S, Hussain M, Mizushima S (1981) Characterization of new membrane lipoproteins and their precursors of *Escherichia coli*. J Biol Chem 256:3125–3129

Ichihara S, Hussain M, Mizushima S (1982) Mechanism of export of outer membrane lipoproteins through the cytoplasmic membrane in *Escherichia coli*. J Biol Chem 257:495–500

Ichihara S, Beppu N, Mizushima S (1984) Protease IV, a cytoplasmic membrane protein of *Escherichia coli*, has signal peptide peptidase activity. J Biol Chem 259:9853–9857

Innis MA, Tokunaga M, Williams ME, Loranger JM, Chang SY, Chang S, Wu HC (1984) Nucleotide sequence of the *Escherichia coli* prolipoprotein signal peptidase (*lsp*) gene. Proc Natl Acad Sci USA 81:3708–3712

Inouye H, Beckwith J (1977) Synthesis and processing of an *Escherichia coli* alkaline phosphatase precursor in vitro. Proc Natl Acad Sci USA 74:1440–1444

Inouye S, Wang SS, Sekizawa J, Halegoua S, Inouye M (1977) Amino acid sequence for the peptide extension on the prolipoprotein of the *Escherichia coli* outer membrane. Proc Natl Acad Sci USA 74:1004–1008

Inouye S, Soberson X, Franceschini T, Nakamura K, Itakura K, Inouye M (1982) Role of positive charge on the amino-terminal region of the signal peptide in protein secretion across the membrane. Proc Natl Acad Sci USA 79:3438–3441

Inouye S, Vlasuk GP, Hsiung H, Inouye M (1984) Effects of mutations at glycine residues in the hydrophobic region of the *Escherichia coli* prolipoprotein signal peptide on the secretion across the membrane. J Biol Chem 259:3729–3733

Inukai M, Inouye M (1983) Association of the prolipoprotein accumulated in the presence of globomycin with the outer membrane of *Escherichia coli*. Eur J Biochem 130:27–32

Inukai M, Takeuchi M, Shimizu K, Arai M (1978) Mechanism of action of globomycin. J Antibiot 31:1203–1205

Ito K (1982) Purification of the precursor form of maltose-binding protein, a periplasmic protein of *Escherichia coli*. J. Biol Chem 257:9895–9897

Ito K, Bassford PJ Jr, Beckwith J (1981) Protein localization in *E. coli*: is there a common step in the secretion of periplasmic and outer-membrane proteins? Cell 24:707–717

Jackson RC (1983) Quantiative assay for signal peptidase. Meth Enzymol 96:784–794

Jackson RC, Blobel G (1977) Post-translational cleavage of presecretory proteins with an extract of rough microsomes from dog pancreas containing signal peptidase activity. Proc Natl Acad Sci USA 74:5598–5602

Jackson RC, Blobel G (1980) Post-translational processing of full-length presecretory proteins with canine pancreatic signal peptidase. Annals New York Acad Sci 343:391–404

Jackson RC, White WR (1981) Phospholipid is required for the processing of presecretory proteins by detergent-solubilized canine pancreatic signal peptidase. J Biol Chem 256:2545–2550

Kamio Y, Lin CK, Regue M, Wu HC (1985) Characterization of the *ileS-lsp* operon in *Escherichia coli*. J Biol Chem 260:5616–5620

Koren R, Burstein Y, Soreq H (1983) Synthetic leader peptide modulates secretion of proteins from microinjected *Xenopus* oocytes. Proc Natl Acad Sci USA 80:7205–7209

Koshland D, Sauer RT, Botstein D (1982) Diverse effects of mutations in the signal sequence on the secretion of *β*-lactamase in *Salmonella typhimurium*. Cell 30:903–914

Kumamoto CA, Oliver DB, Beckwith J (1984) Signal sequence mutations disrupt feedback between secretion of an exported protein and its synthesis in *E. coli*. Nature 308:863–864

Lazdunski C, Baty D, Pages JM (1979) Procaine, a local anesthetic interacting with the cell membrane, inhibits the processing of precursor forms of periplasmic proteins in *Escherichia coli*. Eur J Biochem 96:49–57

Lin JJC, Kanazawa H, Ozols J, Wu HC (1978) An *Escherichia coli* mutant with an amino acid alteration within the signal sequence of outer membrane prolipoprotein. Proc Natl Acad Sci USA 75:4891–4895

Lin JJC, Kanazawa H, Wu HC (1980a) Assembly of outer membrane lipoprotein in an *Escherichia coli* mutant with a single amino acid replacement within the signal sequence of prolipoprotein. J Bacteriol 141:550–557

Lin JJC, Kanazawa H, Wu HC (1980b) Purification and characterization of the outer membrane lipoprotein from an *Escherichia coli* mutant altered in the signal sequence of prolipoprotein. J Biol Chem 255:1160–1163

Lively MO, Walsh KA (1983) Hen oviduct signal peptidase is an integral membrane protein. J Biol Chem 258:9488–9495

Marchal C, Perrin D, Hedgpeth J, Hofnung M (1980) Synthesis and maturation of lambda receptor in *Escherichia coli* K-12: in vivo and in vitro expression of gene lamB under lac promoter control. Proc Natl Acad Sci USA 77:1491–1495

McAda PC, Douglas MG (1982) A neutral metallo-endoprotease involved in the processing of an F1-ATPase subunit precursor in mitochondria. J Biol Chem 257:3177–3182

Michaelis S, Beckwith J (1982) Mechanism of incorporation of cell envelope proteins in *Escherichia coli*. Ann Rev Microbiol 36:435–464

Minkley EG Jr (1984) Purification and characterization of pro-TraTp, the signal sequence containing precursor of a secreted protein encoded by the sex factor. J Bacteriol 138:464–473

Mollay C, Vilas U, Kreil G (1982) Cleavage of honeybee prepromelittin by an endoprotease from rat liver microsomes: identification of intact signal peptide. Proc Natl Acad Sci USA 79:2260–2263

Novak P, Ray PH, Dev IK (1985) Localization and purification of two signal peptide hydrolases from *Escherichia coli*. J Biol Chem 261:420–427

Omoto S, Ogino H, Inouye S (1981) Studies on SF-1902 A_2–A_5, minor components of SF-1902 (Globomycin). J Antibiot 34:1416–1423

Pacaud M (1982) Purification and characterization of two novel proteolytic enzymes in membranes of *Escherichia coli*. J Biol Chem 257:4333–4339

Perlman D, Halvorson HO (1983) A putative signal peptidase recognition site and sequence in eukaryotic and prokaryotic signal peptides. J Mol Biol 167:391–409

Pugsley AP, Schwartz M (1985) Export and secretion of proteins by bacteria. FEMS Microbiol Rev 32:1–36

Randall LL, Hardy SJS (1984) Export of protein in bacteria. Microbiol Rev 48:290–298

Ray PH, Dev IK, Miller TA, Bassford PJ Jr (1986) Evidence for another signal peptidase: In vitro processing of maltose-binding protein precursors having alterations in their signal sequences. Submitted for publication

Regue M, Remenick J, Tokunaga M, Mackie GA, Wu HC (1984) Mapping of the lipoprotein signal peptidase gene (*lsp*). J Bacteriol 158:632–635

Roggenkamp R, Kunstermann-Kuhn B, Hollenberg CP (1981) Expression and processing of bacterial β-lactamase in the yeast *Saccharomyces cerevisiae*. Proc Natl Acad Sci USA 78:4466–4470

Roggenkamp R, Hoppe J, Hollenberg CP (1983) Specific processing of the bacterial β-lactamase precursor in *Saccharomyces cerevisiae*. J Cell Biochem 22:141–149

Roggenkamp R, Dargatz H, Hollenberg CP (1985) Precursor of β-lactamase is enzymatically inactive. J Biol Chem 260:1508–1512

Shapiro AL, Vinuela E, Maizel JV Jr (1967) Molecular weight estimation of polypeptide chains by electrophoresis in SDS-polyacrylamide gels. Biochem Biophys Res Comm 28:815–820

Silhavy TJ, Shuman HA, Beckwith J, Schwartz M (1977) Use of gene fusions to study outer membrane protein localization in *Escherichia coli*. Proc Natl Acad Sci USA 74:5411–5415

Silhavy TJ, Bassford PJ Jr, Beckwith JR (1979) A genetic approach to the study of protein localization in *Escherichia coli*. In: Inouye M (ed) Bacterial outer membranes: biogenesis and functions. Wiley, New York, pp 203–254

Silhavy TJ, Benson SA, Emr SD (1983) Mechanisms of protein localization. Microbiol Rev 47:313–344

Silver P, Wickner W (1983) Genetic mapping of the *Escherichia coli* leader (signal) peptidase gene (*lep*): a new approach for determining the map position of a cloned gene. J Bacteriol 154:569–572

Silver P, Watts C, Wickner W (1981) Membrane assembly from purified components. I. Isolated M13 procoat does not require ribosomes or soluble proteins for processing by membranes. Cell 25:341–345

Stern JB, Jackson RC (1985) Peptide products of the cleavage of bovine preprolactin by signal peptidase. Arch Biochem Biophys 237:244–252

Strauss AW, Zimmerman M, Boime I, Ashe B, Mumford RA, Alberts AW (1979) Characterization of an endopeptidase involved in pre-protein processing. Proc Natl Acad Sci USA 76:4225–4229

Strauss AW, Zimmerman M, Mumford RA, Alberts AW (1980) Processing of pre-proalbumin and pre-placental lactogen. Ann New York Acad Sci 343:168–191

Talmadge K, Stahl S, Gilbert W (1980) Eukaryotic signal sequence transports insulin antigen in *Escherichia coli*. Proc Natl Acad Sci USA 77:3369–3373

Tokunaga H, Wu HC (1984) Studies on the modification and processing of prolipoprotein in *Escherichia coli*. J Biol Chem 259:6098–6104

Tokunaga M, Loranger JM, Wolfe PB, Wu HC (1982a) Prolipoprotein signal peptidase in *Escherichia coli* is distinct from the M13 procoat protein signal peptidase. J Biol Chem 257:9922–9925

Tokunaga M, Tokunaga H, Wu HC (1982b) Post-translational modification and processing of *Escherichia coli* prolipoprotein in vitro. Proc Natl Acad Sci USA 79:2255–2259

Tokunaga M, Loranger JM, Wu HC (1983) Prolipoprotein modification and processing enzymes in *Escherichia coli*. J Biol Chem 259:3825–3830

Tokunaga M, Loranger JM, Wu HC (1984) A distinct signal peptidase for prolipoprotein in *Escherichia coli*. J Cell Biochem 24:113–120

Tokunaga M, Loranger JM, Chang S-Y, Regue M, Chang S, Wu HC (1985) Identification of prolipoprotein signal peptidase and genomic organization of the *lsp* gene in *Eschericia coli*. J Biol Chem 260:5610–5615

Vimr ER, Green L, Miller CG (1983) Oligopeptidase-deficient mutants of *Salmonella typhimurium*. J Bacteriol 153:1259–1265

Vlasuk GP, Inouye S, Ito H, Itakura K, Inouye M (1983) Effects of the complete removal of basic amino acid residues from the signal peptide on secretion of lipoprotein in *Escherichia coli*. J Biol Chem 258:7141–7148

Vlasuk GP, Inouye S, Inouye M (1984) Effects of replacing serine and threonine residues within the signal peptide on the secretion of the major outer membrane lipoprotein of *Escherichia coli*. J Biol Chem 259:6195–6200

von Heijne G (1984) How signal sequences maintain cleavage specificity. J Mol Biol 173:243–251

Watson MEE (1984) Compilation of published signal sequences. Nucleic Acids Research 12:5145–5164

Watts C, Wickner W, Zimmermann R (1983) M13 procoat and a pre-immunoglobulin share processing specificity but use different membrane receptor mechanisms. Proc Natl Acad Sci USA 80:2809–2813

Wickner W, Mandel G, Zwizinski C, Bates M, Killick T (1978) Synthesis of phage M13 coat protein and its assembly into membranes in vitro. Proc Natl Acad Sci USA 75:1754–1758

Wolfe PB, Silver P, Wickner W (1982) The isolation of homogeneous leader peptidase from a strain of *Escherichia coli* which overproduces the enzyme. J Biol Chem 257:7898–7902

Wolfe PB, Wickner W, Goodman JM (1983a) Sequence of the leader peptidase gene of *Escherichia coli* and the orientation of leader peptidase in the bacterial envelope. J Biol Chem 258:12073–12080

Wolfe PB, Zwizinski C, Wickner W (1983b) Purification and characterization of leader peptidase from *Escherichia coli*. Methods Enzymol 97:40–46

Wolfe PR, Rice M, Wickner W (1985) Effects of two *sec* genes on protein assembly into the plasma membrane of *Escherichia coli*. J Biol Chem 260:1836–1841

Yamagata H, Daishima K, Mizushima S (1983a) Cloning and expression of a gene coding for the prolipoprotein signal peptidase of *Escherichia coli*. FEBS Lett 158:301–303

Yamagata H, Taguchi N, Daishima K, Mizushima S (1983b) Genetic characterization of a gene for prolipoprotein signal peptidase in *Escherichia coli*. MGG 192:10–14

Yu F, Yamada H, Daishima K, Mizushima S (1984) Nucleotide sequence of the *lspA* gene, the structural gene for lipoprotein signal peptidase of *Escherichia coli*. FEBS Lett 173:264–268

Zwizinski C, Wickner W (1980) Puficiation and characterization of leader (signal) peptidase from *Escherichia coli*. J Biol Chem 255:7973–7977

Zwizinski C, Date T, Wickner W (1981) Leader peptidase is found in both the inner and outer membranes of *Escherichia coli*. J Biol Chem 256:3593–3597

Protein Secretion in Bacilli

M. SARVAS

1 Introduction

Bacilli and especially *Bacillus subtilis* have recently emerged as model organisms worthy of consideration when studying the molecular mechanisms of bacterial protein export. Several factors contribute to this. Most species of *Bacillus* have a high capacity to secrete many proteins, indicating a developed machinery of protein secretion that can be expected to be amenable to elucidation. Further technical advantage is provided by the structure of the gram-positive cell envelope, because of which exported proteins need to traverse only one membrane (the cytoplasmic membrane) to be exposed outside and finally released from the cell. In this respect protein secretion by gram-positive bacteria can even be considered analogous to that of eukaryotic cells, and may be different from that of *Escherichia coli* in an important way. Although *E. coli* has also served successfully as a model organism for research on protein secretion (SILHAVY et al. 1983), the complicated structure of its cell envelope and the lack of true protein secretion across its two membranes are disadvantages compensated only by the enormous amount of basic knowledge of its cellular physiology, structure, and genetics.

Many proteins secreted from Bacilli are exoenzymes of industrial importance (for a review see DEBABOV 1982; PRIEST 1977; GLENN 1976). This, together

National Public Health Institute, Mannerheimintie 166, SF-00280 Helsinki 28

Current Topics in Microbiology and Immunology, Vol. 125

with the recently increased interest in biotechnology, has led to considerable research actitivity in the field. Most importantly, this has stimulated the development and utilization of recombinant-DNA techniques applicable to Bacilli. The main goals have been to improve the production of both established and new types of bacillar enzymes, and to develop this well-known and harmless industrial organism to a host for the production of foreign proteins. Although much of this research has been decidedly application oriented, it has contributed significantly to the development of bacillar molecular biology, including the molecular mechanism of secretion.

2 Types of Proteins Exported from Bacilli

A large number of different types of proteins are located outside the cytoplasmic membrane in bacilli. All these are exported proteins in the sense that they must traverse the cytoplasmic membrane in order to reach their final destination. However, they fall into several categories by their nature and exact location.

The best known are a number of "true" soluble exoproteins – mainly exoenzymes – secreted often in large quantities, out of the cell and into the growth medium, very little if any remaining cell-associated. Typical examples of these proteins are extracellular degradative enzymes such as α-amylases and proteases (for review, see PRIEST 1977; DEBABOV 1982).

The rest are a heterogeneous group of proteins, which are at least to some extent associated with the cell wall and the cytoplasmic membrane. Some of these proteins are clearly located outside the cytoplasmic membrane – for example, autolytic enzymes (SHOCKMAN and BARRETT 1983), peptidoglycan-associated proteins (the best-known example of this kind of protein is the staphylococcal protein A, SJÖDAHL 1977; UHLEN et al. 1984), S-layer proteins (SLEYTR and MESSNER 1983), and the hydrophilic protein domain of bacillar lipoproteins (see below). In many cases their exact location is ill defined. Under certain growth conditions they may also be secreted or released into the growth medium, blurring to some extent the distinction between the main types of exported proteins.

Many different types of these proteins have been used for studying protein secretion of bacilli. The emphasis has been on exoproteins secreted in abundance, but many laboratories have also chosen other proteins as models, which have some typical characteristics of interest. Thus there has been a rich source of distinctly different proteins and experimental systems to contribute to the emerging pattern of protein secretion. On the other hand this variety also complicates efforts to build up a comprehensive model of secretion mechanism, or to find out whether there is more than one secretory pathway for different types of proteins.

3 Protein Translocation Across the Cytoplasmic Membrane in Bacilli

Compared with the rather advanced state of knowledge concerning the mechanisms of eukaryotic protein secretion, and the recent rapid progress in the best-known prokaryotic cell, *E. coli* (see e.g., PUGSLEY and SCHWARTZ 1985; SILHAVY et al. 1983; Müller and Blobel; Beckwith and Ferro-Novick; this volume), much less is known of these aspects in bacilli. Serious obstacles for progress have been the lack of a fully effective in vitro system for studying the process and the relative lack of basic genetic data in bacilli. Consequently, most of the knowledge on the mechanism of protein secretion of bacilli and its components is so far indirect.

Since the demonstration of in vitro synthesis of exported proteins of *E. coli* by membrane-bound ribosomes in crude membrane preparations (CANCEDDA and SCHLESSINGER 1974; RANDALL and HARDY 1977), a comparable approach has been used to study the synthesis of secreted proteins of bacilli. Two exoproteins of *B. licheniformis* (α-amylase and β-glucosidase) were found to be exclusively synthesized by membrane-bound ribosomes (THIRUNAVUKKARASU and PRIEST 1983). In more elaborate experiments, SMITH and coworkers applied an external label to proteins during their synthesis and export through the cytoplasmic membrane of *B. subtilis* (SMITH et al. 1978, 1979, 1981). A significant proportion of the labeled polypeptides was found to be associated with polysomes, suggesting that the nascent polypeptides were translocated across the cytoplasmic membrane while still attached to ribosomes and not yet fully elongated. Among the labeled polypeptides specific exported proteins were identified immunologically. These data indicate that secreted bacillar proteins are synthesized, like similar proteins in eukaryotes, by membrane-bound ribosomes and translocated across the cytoplasmic membrane before the completion of their elongation.

However, experiments performed so far in bacilli cannot distinguish between the two proposed modes of translocation, either strict cotranslational translocation as shown to take place in vesicles of the rough endoplasmic reticulum of eukaryotic cells (see Müller and Blobel, this volume) or post-translational translocation as originally suggested by Wickner for the export of proteins in *E. coli* (see Wickner, this volume). The latter hypothesis has recently been supported in *E. coli* both by experiments by RANDALL (RANDALL 1983), showing translocation of periplasmic proteins across the cytoplasmic membrane only after a substantial portion of the polypeptide has been synthesized, and by the direct demonstration of post-translational insertion of periplasmic and outer membrane proteins into inverted cytoplasmic membrane vesicles (Tai; Müller and Blobel; this volume). A post-translational mode of export and translocation has been suggested for alkaline phosphatase in *B. subtilis* based on the presence of substantial amounts of the protein in the cytoplasm and its possible later release into the growth medium. However, the protein has not been characterized well enough on a molecular level to provide direct evidence for its exact mode of translocation (see below for a more detailed discussion).

4 Cellular Components of the Protein Export Machinery of Bacilli

An important question in studies of bacterial protein export has been to what extent its complexity and cellular components are similar to those of eukaryotic cells. Essential elements of the system in eukaryotes are a cytoplasmic signal recognition particle (SPR) and integral membrane proteins that bind ribosomes to the membrane (ribophorins) and initiate protein translation once the SPR and ribosomes are bound to the membrane (docking protein) (see Müller and Blobel, this volume). Genetic studies in *E. coli* have already revealed several genes and proteins involved in protein secretion (for a review, see PUGSLEY and SCHWARTZ 1955; SILHAVY et al. 1983) and suggest a complexity comparable to that of the eukaryotic system.

Studies of proteins associated with membrane-ribosome complexes have recently provided circumstantial evidence for the presence of several export-associated proteins also in *B. subtilis*. MARTY-MAZARS et al. (MARTY-MAZARS et al. 1983; HORIUCHI et al. 1983a, b) have developed a fractionation procedure to separate membrane-ribosome complexes from less dense ribosome-free cytoplasmic membrane vesicles of *B. subtilis*. The membrane fraction associated with the ribosomes was found to contain up to six proteins, which were almost absent from the rest of the membrane. They were immunologically distinct and their quantitation suggested that these proteins were present in fixed ratios in a complex. The most interesting of these proteins is a 64-kD protein. Initially it was found to be associated with membrane-ribosome complexes, where it is situated on the inner surface of the cytoplasmic membrane and covered by ribosomes, strongly suggesting a role in protein secretion (HORIUCHI et al. 1983b). Substantial amounts of the protein are also found as free protein in the cytoplasm (HORIUCHI et al. 1983a, b; CAULFIELD et al. 1984). Most importantly, it has also been shown to be a component of the so called S-complex (consisting of four proteins: 64, 60, 41, and 36 kD), which is rather surprisingly associated exclusively with free monosomes and (presumably also) polysomes. Only the 64 kD protein was also found in membrane-ribosome complexes (CAULFIELD et al. 1984).

ADLER and ARVIDSON (ADLER and ARVIDSON 1984a) have detected in *Staphylococcus aureus* a 60 kD protein, which is also specifically associated with membrane-bound ribosomes and released with mild detergent from protoplasts. This protein – like the 64 kD protein in bacilli – was found also in the cytoplasm. Antibodies against the 64 kD staphylococcal protein precipitate a complex of four proteins from detergent-solubilized membranes and from the cytoplasm of *S. aureus* (ADLER and ARVIDSON 1984b). Most interestingly, the antibody also immunoprecipitates a 64 kD protein and a set of four proteins very similar to the above-descrived S-complex from solubilized protoplasts of *B. subtilis* (ADLER and ARVIDSON 1984b).

Although the proteins of *B. subtilis* and *S. aureus* discussed above have not been proven to have a function in protein secretion, their association with membrane-ribosome complexes makes such a role very probable. Furthermore, the protein complexes found have a clear resemblance to the eukaryotic signal recognition particle. The findings of CAULFIELD et al. (1984) also suggest a cyclic role for the 64 kD protein in the initiation of the process of secretion and attachment of ribosomes to the cytoplasmic membrane. If so, this would extend the analogy between eukaryotic and prokaryotic secretory mechanisms to functional aspects of the machinery.

One potential advantage in studying the mechanism of protein secretion in bacilli is the availability of relatively simple screening methods to look for mutants with an altered capacity to secrete proteins. Due to the structure of its cell wall, *E. coli* has been a problematic model organism in this respect (see SILHAVY et al. 1983). Indeed, numerous mutants with a decreased or increased level of various exoenzymes have been described in bacilli, especially *B. subtilis*. Many such mutants have been reviewed by PRIEST (1977). The majority of these mutations are pleiotropic, affecting several exoenzymes and also other properties of the cell. The characterization of such mutants is complicated by the fact that the synthesis of many exoenzymes may be under a common control, most notably affected by sporulation (see PRIEST 1977). Nevertheless, one could also expect to find among them mutants in which a gene determining components of the secretory machinery have been altered. Among the mutations so far characterized, those in the *sacU* gene of *B. subtilis* seem to fall into this category.

Two sets of mutations mapping in the *sacU* gene have been isolated. The $sacU^-$ mutations decrease the amount of extracellular levansucrase and proteases (STEINMETZ et al. 1976; KUNST et al. 1974). The more interesting set of mutations, designated $sacU^h$, have many pleiotropic effects, including overproduction of extracellular levansucrase, proteases, α-amylase, and possibly other exoenzymes, lack of flagella, and poor transformability (STEINMETZ et al. 1976; YONEDA and MARUO 1975; YONEDA et al. 1973; AYUSAWA et al. 1975). The (wild-type) $sacU^+$ gene has recently been cloned in both *B. subtilis* and *E. coli* and shown to encode a 46 kD protein (AUBERT et al. 1985). A protein of the same size was found in the membrane fraction of *B. subtilis*; this protein was increased in amount in the pleiotropic hyperproducing $SacU^h$ mutants and virtually absent in $SacU^-$ mutants. The $sacU^h$ mutation was furthermore shown not to increase the amount of the mRNA of one exoenzyme, the levansucrase. It thus appears probable that the increased rate of production of several exoenzymes in $SacU^h$ mutants is caused by the increased amount of the membrane-associated 46 kD protein. This is therefore also a possible component of the secretory machinery.

5 Signal Peptides of Bacillar Exoproteins

The most universal feature of protein secretion is the synthesis of most secretory protein as precursors with *N*-terminal signal peptides. This is a rule in eukaryotes shared to a large extent also in *E. coli*. Several genes coding for exoenzymes of bacilli have now been cloned and many of them sequenced (Table 1), providing information on the presence and character of signal peptides. The known amino acid sequence of the secreted enzyme can in most cases be compared

Table 1. Signal peptides of exoproteins of bacilli

Exoprotein	Species	N-Terminal amino acid sequence of the primary translate	Reference
α-Amylase	*B. amyloliquefaciens*	MIQKRKRTVSFRLVLMCT-LLFVSLPITKTSA.VN	Palva et al. (1981)
α-Amylase	*B. subtilis*	MFAKRFKTSLLPLFAGF-LLLFYLVLAGPAAASAET	Yang et al. (1983), Yamazaki et al. (1983), Ohmura et al. (1984)
α-Amylase	*B. licheniformis*	MKQHKRLYARLLTLLFAL-IFLLPHSAAAA.AN	Stephens et al. (1984), Sibakov and Palva (1984)
Levansucrase	*B. subtilis*	MNIKKFAKQATILTFTTA-LLAGGATQAFA.KE	Fouet et al. (1984)
β-Glucanase	*B. subtilis*	MPYLKRVLLLLVTGLFMS-LFAVTATASAKT	Murphy et al. (1984)
Xylanase	*B. pumilis*	MNLRKLRLLFVMCIGLTLI-LTAVPAHA.RT	Fukusaki et al. (1984)
β-Lactamase	*B. cereus*	MMILKNKRMLKIGICVGI-LGLSITSLEAFTGESLQ	Lampen et al. (1984)
Serine protease	*B. subtilis*	MRSKKLTISLLFALTLIFT-MAFSNMSAQAAGKSSTEK	Wong et al. (1984)
Subtilisin	*B. amyloliquefaciens*	MRGKKVTISLLFALALIFT-MAFGSTSSAQAAGKSNG	Wells et al. (1983)
Neutral protease	*B. subtilis*	MGLGKKLSVRVAASFMSL-SISLPGVQAAEG	Yang et al. (1984)
β-Lactamase	*B. licheniformis*	MKLWFSTLKLKKAAAVLL-FSCVALAG.CA	Kroyer and Chang (1981), Neugebauer et al. (1981)

. indicates a known cleavage site of the signal peptide

with the deduced sequence of the primary translate, showing the presence of an amino-terminal peptide extension in the latter. According to this criterion all the exoenzymes studied so far are obviously made as precursors with a cleavable signal peptide or prepeptide.

The overall pattern of the N-terminal part of the primary translate of all of these proteins conforms rather closely with the general pattern of cleavable prokaryotic signal peptides (Watson 1984). There is invariably an N-terminal hydrophilic and basic sequence followed by a stretch of uncharged residues (the hydrophobic core). No evident sequence homology between the signal peptides of different exoproteins has been detected in any species studied so far. A demonstative case is the signal peptides of α-amylases from *B. amyloliquefaciens* and *B. licheniformis*. The mature proteins have considerable sequence homology between them, but their signal peptides are no more homologous than signal peptides in general (Sibakov and Palva 1984; Stephens et al. 1984).

However, the bacillar signal peptides also have some distinctive features not frequently found in other signal peptides. The average length of the cleavable peptide is long, usually between 30 and 40 residues (Table 1), as compared with an average of 22 in other species. Long signal peptides may be a general feature of gram-positive bacteria, because those known in *Staphylococcus* also seem to be in the range of 30–40 residues (LÖFDAHL et al. 1983; SAKO and TSUCHIDA 1983). There also is a trend toward a rather long noncharged stretch (14–23 residues in well-characterized sequences). The N-terminal hydrophilic segment is often strikingly rich in basic residues. Up to five arginines and lysines are found in the precursor of α-amylase of *B. amyloliquefaciens* (PALVA et al. 1981) and the average number of charged residues is about four.

It is so far unclear whether the described unusual features of the primary structure have any functional importance. The thick cell wall outside the cytoplasmic membrane in bacilli and other gram-positive bacteria could set unique requirements on their secretory machinery, reflected also in the structure of their signal peptides. However, so far there is no experimental evidence for such a relationship. In a few studies structural changes have been introduced by site-directed mutagenesis in the α-amylase gene of *B. amyloliquefaciens* (PALVA et al. 1982) or *B. subtilis* (OHMURA et al. 1984a). The changes, however, have affected such C-terminal parts of the signal peptides that have no distinctive features when compared with signal peptides in general. In these studies a deletion removed a short stretch of residues, including the site of cleavage of the signal peptides and a few C-terminal residues of the hydrophobic core. In agreement with the effects of analogous deletions in signal peptides in *E. coli* (SILHAVY et al. 1983), the deletion prevented the cleavage of the signal peptide from the (defective) precursor (PALVA et al. 1982; OHMURA et al. 1984a), without preventing the synthesis of these proteins or their association with the cytoplasmic membrane.

The expression of several cloned bacillar exoprotein genes has been studied in *E. coli* (Table 2). In most cases the protein was found mainly or partially as a soluble protein in the periplasmic space or there were other indications of its translocation across the cytoplasmic membrane and/or of signal peptide processing. In spite of their slightly unusual features, bacillar signal peptides can therefore function in *E. coli*, and like many other heterologous secretory proteins, bacillar exoproteins are translocated across a heterologous cytoplasmic membrane without difficulty.

The amino acid sequence at the cleavage site of the bacillar signal peptides also seems to conform with the general pattern in prokaryotes. In all those cases where proper analytical data are available, the cleavage seems to take place after an alanine residue or, after a glycine residue in lipoproteins (PALVA et al. 1981; STEPHENS et al. 1984; SIBAKOV and PALVA 1984; FOUET et al. 1984b; FUKUSAKI et al. 1984; KROYER and CHANG 1981; FISHMAN et al. 1980; Wu and Tokunaga, this volume). However, there may be some ambiguity in deducing the cleavage site of the signal peptide from protein. The protein may undergo further proteolytic digestion after the cleavage of the signal peptide. A stepwise cleavage of more than one N-terminal oligopeptide from the penicillinase of *B. licheniformis* after the cleavage of the signal peptide has been well documented, at least one of them taking place in the culture medium (IZUI et al. 1980; SMITH et al. 1981). The ragged N-terminus of some secreted bacillar exoproteins (PRIEST 1977) suggests that nibbling in the culture medium may occur commonly. Furthermore, the amino acid sequence of the precursor (deduced from the DNA

Table 2. Genes of bacillar exoproteins cloned in *E. coli*

Source	Protein	Localization in *E. coli*	Reference
B. subtilis	α-Amylase	(Deleterious)	YANG et al. (1983) OHMURA et al. (1983)
B. subtilis	Levansucrase	Extracellular (?)	GAY et al. (1983), FOUET et al. (1984a, b)
B. subtilis	β-Glucanase	ND	CANTWELL and MCCONNELL (1983)
B. licheniformis	α-Amylase	Periplasmic	SIBAKOV and PLAVA (1984), JOYET et al. (1984)
B. licheniformis	Alkaline phosphatase	ND	HULETT (1984)
B. coagulans	α-Amylase	Periplasmic	CORNELIS et al. (1982), WILLEMOT and CORNELIS (1983)
B. cereus	β-Lactamase type I	Periplasmic membrane bound	SLOMA and GROSS (1983)
B. cereus	Cereolysin	ND	KREFT et al. (1983)
B. pumilis	Xylanase	Cytoplasmic	FUKUSAKI et al. (1984) PANBANGRED et al. (1983)
B. stearothermophilus	α-Amylase	Periplasmic	TSUKAGOSHI et al. (1984a)
Bacillus sp.	Xylanase A	Periplasmic extracellular	HONDA et al. (1985)

ND, not determined

sequence) preceding the site of the N-terminus of some mature exoproteins, like the α-amylase (YANG et al. 1983; YAMAZAKI et al. 1983; OHMURA et al. 1984a) or β-glucanase (MURPHY et al. 1984) of *B. subtilis* is not "typical" of a cleavage site (no alanine, glycine, or other small residue preceding the N-terminus). This suggests that the primary cleavage of the signal peptide has taken place at a probable cleavage site identifiable in each case, followed by (a) further cleavage(s). It is not clear whether the proteases involved are specific ones or exoproteases secreted by most bacillar species, especially in the stationary phase of cultures.

6 Propeptides of Bacillar Exoproteases

Recent cloning and sequencing of the two major types of secreted proteases, the alkaline (serine) protease and the neutral (metallo) protease of *B. subtilis* (WONG et al. 1984; YANG et al. 1984) and *B. amyloliquefaciens* (WELLS et al. 1983; VASANTHA et al. 1984) have revealed a novel feature of bacterial secretory proteins. The primary translates of these genes have a very large N-terminal extension – a propeptide of 107, 106, and 222 residues, respectively – when compared with the secreted form of the proteins. The signal peptide is unlikely to comprise this whole stretch of residues, especially because the first 30 or so N-terminal residues of each primary translate have all the characteristic features of a typical signal peptide, followed by a putative cleavage site.

It seems more likely that the propeptide proteins have some other function or role than ordinary signal peptides. As pointed out in the papers referred to above, these peptides are analogous to propeptides of eukaryotic secretory protease. They also share some structural features with them, such as a high content of clustered charged residues.

The function of the propeptide is unknown. Its presence in all bacillar proteases characterized so far and its absence in all other exoproteins (Table 1) suggests that it may not have a role in the translocation of these protein across the cytoplasmic membrane. In a way analogous to eukaryotic proteases the propeptide may render the protease inactive before it is needed, e.g., inactive close to some protease-sensitive components of the cell membrane and the cell wall, to be activated later, possibly by an autocatalytic process (WELLS et al. 1983; VASANTHA et al. 1984). It has also been suggested that the propeptide might serve, by virtue of its net positive charge, to bind the protein to the cell membrane (VASANTHA et al. 1984). Clearly the function of the propeptide or the cellular process in which it is involved can only be elucidated after it has been better characterized, in respect to its possible enzymatic activity, cellular location, and mode of processing.

7 Lipoproteins of Bacilli

Among the exported bacillar proteins a unique class of proteins remain after their translocation attached to the outer surface of the membrane by a lipid moiety covalently bound to the protein. The large extramembraneous domain of these proteins is often detached proteolytically from the membrane to a variable extent and released into the growth medium as an enzymatically active exoprotein. Especially in the stationary phase of the culture, most of the protein – like conventional secretory proteins – may be found in the growth medium. Indeed, the best-characterized example of these proteins, the penicillinase of *B. licheniformis*, has been extensively studied as a model exoprotein to elucidate the molecular mechanism of protein secretion from bacilli (for review, see LAMPEN et al. 1984; Wu and Tokunaga, this volume).

The lipid moiety of these proteins has a structure which is uncommon in nature and previously found only in the lipoproteins of the cell envelope of gram-negative bacteria (HANTKE and BRAUN 1973; ICHIHARA et al. 1981). In short, these proteins are lipophilic because of the covalent modification of their N-terminal cysteine residue with glyceride thioether and with an amide-linked fatty acid moiety. The lipid moiety of all lipoproteins seems to be identical. Despite the fundamental differences between the structure of the cell envelope of gram-negative and gram-positive bacteria, the structural similarity of the lipoproteins is extensive, including a similar mode of synthesis, translocation across the cytoplasmic membrane, and unique steps of processing and modification. These aspects of lipoproteins are discussed on detail by Wu and Tokunaga, in this volume.

Although the structure and location of the bacillar lipoproteins on the outer surface of the cytoplasmic membrane has been established, their physiological role has remained less clear, especially the function of the membrane-bound form in relation to the apparent exoprotein released. The two bacillar lipoproteins so far fully characterized are β-lactamases – the penicillinase of *B. licheniformis* (NIELSEN et al. 1981; SLOMA and GROSS 1983; SMITH et al. 1981) and the β-lactamase III of *B. cereus* (NIELSEN and LAMPEN 1982b). The presence of additional lipoproteins in *B. licheniformis* is suggested by the association of fatty acids and glycerol with some uncharacterized proteins of the cytoplasmic membrane (NIELSEN and LAMPEN 1982b; SMITH et al. 1981). Furthermore, a β-lactamase of another gram-positive genus, the penicillinase of *Staphylococcus aureus*, has a membrane-bound form of similar lipoprotein nature (NIELSEN and LAMPEN 1982a).

Binding of these enzymes to the cytoplasmic membrane might serve the purpose of retaining these enzymes in the vicinity of the cell in order to enhance its resistance to β-lactam antibiotics (NIELSEN and LAMPEN 1982b; CHANG et al. 1982a). Alternatively, the lipoproteins may be intermediates in a secretory pathway specific to certain group(s) of exoproteins. The kinetics of release of the exopenicillinase of *B. licheniformis* at different pH suggest that at least some of the exopenicillinase is released proteolytically from the membrane-bound lipoprotein (CRANE et al. 1973; LAMPEN and NIELSEN 1982). This might be accomplished by some of the exoproteases of broad substrate specificity that are synthesized by *B. licheniformis* in abundance, but the presence of a specific cleavage enzyme has also been suggested (TRAFICANTE and LAMPEN 1977; SMITH et al. 1981). When the *penP* gene coding for the penicillinase of *B. licheniformis* was cloned in *B. subtilis* (IMANAKA et al. 1981, 1983), in *B. strearothermophilus* (FUJII et al. 1982), and in a thermophilic *Bacillus* sp. (IMANAKA 1983), the proportion of exopenicillinase was only 5%–15% of the total, compared with about 50% released in *B. licheniformis*. An explanation for the decreased release could be a lack of a proper (specific?) proteolytic enzyme.

On the other hand, the formation of the lipoprotein does not seem to be an absolute requirement for the export and release of the penicillinase of *B. licheniformis* in soluble form. Using site-specific mutagenization, MÉZES et al. (1983) have constructed a *penP* gene, which lacks the last four codons of the signal sequence and the 5′ terminal Cys codon of the lipoprotein form (see Table 1). The deleted peptide is crucial for the processing and modification of lipoproteins (see Wu and Tokunaga, this volume). As expected, the penicillinase synthesized in *B. subtilis* as directed by the truncated penicillinase gene was not modified with fatty acids. In spite of this, about 50% of the enzyme was present on the outer surface of the cytoplasmic membrane as judged by its accessibility both to its substrate and to trypsin added exogenously.

In the exponential phase of growth very little mutant penicillinase was released from the cell, while the bulk of it remained cell associated in a form corresponding in size to the unprocessed primary translate of the truncated gene. However, in later phases of the growth the enzyme was quantitively released into the growth medium as a soluble protein processed by N-terminal cleavage(s) to the same size as the exoenzyme found in cultures of *B. lichenifor-*

mis. Cleavage of the signal peptide from the primary translate of the intact *penP* gene has also been shown to take place in vitro without prior modification to lipoprotein (CHANG et al. 1982a). Thus the complete sequence of events resulting in the secretion of penicillinase can take place without the glyceride-lipoprotein intermediate. The late release of the exopenicillinase and the specificity of the cleavage (MÉZES et al. 1983) suggest that the proteases releasing the penicillinase from the unprocessed precursor are exoproteases rather than a specific signal peptidase. It would be of interest to find out whether some normal exoproteins might be released from the precursor by such proteases instead of by the signal peptidase.

8 Possible Membrane-Bound Intermediates of Other Exoenzymes

Whether membrane-bound intermediates, like the glyceride lipoproteins of β-lactamases, are a general feature of protein secretion of bacilli and of other gram-positive bacteria has been an intriguing question. Their presence was suggested by early works in many laboratories, although in most cases the findings have not been backed later by extended studies. There is a general finding that even after extensive washing a small fraction of many well-characterized soluble exoenzymes remains bound to the cell membrane (PRIEST 1977). Comparison of the cell-bound and secreted forms of proteases of *B. subtilis* (MÄNTSÄLÄ and ZALKIN 1980), levansucrase of *B. subtilis* and *B. amyloliquefaciens* (MÄNTSÄLÄ and PUNTALA 1982), and α-amylase of *B. subtilis* (NAGATA et al. 1974) have shown them to be antigenically and/or enzymatically identical and unlikely to be two different proteins. The lipophilic properties of the membrane-bound proteases of *B. subtilis* (MÄNTSÄLÄ and ZALKIN 1979, 1980) and, more circumstantially, the extractability of the membrane-bound forms of levansucrase by phenol (PETIT-GLATRON et al. 1980) seemed to indicate that the membrane-bound fraction represented a special form of the protein and not only exoenzyme trapped in membrane vesicles.

A further method extensively applied to probe for the presence of possible lipoprotein intermediates of secreted proteins has been to inhibit fatty acid synthesis specifically with the antibiotic cerulenin. In agreement with its lipoprotein nature, the synthesis of the penicillinase of *B. licheniformis* was indeed drastically inhibited by cerulenin (FISHMAN et al. 1978). Subsequently, cerulenin has been shown to inhibit the synthesis of several bacillar exoenzymes without affecting protein synthesis in general. The proteins studied include levansucrase (CAULFIELD et al. 1979) of *B. subtilis*, α-amylase, and a protease of *B. amyloliquefaciens* (PATON et al. 1980) as well as the secreted enterotoxin of *Staphylococcus aureus* (ALTENBERN 1977; BERKELEY et al. 1978).

These studies would suggest a close, specific coupling of the synthesis of fatty acids and many secreted proteins of bacilli, suggesting a putative precursor of lipoprotein character. However, kinetic studies on the effect of cerulenin on the synthesis and induction of levansucrase of *B. subtilis* raise doubts of the specificity of its action (PETIT-GLATRON and CHAMBERT 1981). Furthermore, some amphiphilic compounds like dodecanoic acid, phenethyl alcohol, procaine, and ethanol also affect the synthesis of levansucrase (PETIT-GLATRON and CHAMBERT 1981) and several other bacillar exoenzymes, including the penicillinase of *B. licheniformis*, in an apparently specific way. The main effect of these compounds is to disrupt the structure of biological membranes (SILVER and WENDT 1967; FISHMAN et al. 1980), which could account for their effect on the synthesis of exoproteins. Membrane-disrupting compounds also have very similar and again apparently specific effects on the synthesis of exported proteins in *E. coli* (MURGIER et al. 1982; HALEGOUA and INOUYE 1979). For these proteins, the presence of intermediates of lipoprotein character is virtually excluded (see e.g., SILHAVY et al. 1983).

The membrane-bound forms of none of the above enzymes have been characterized beyond their lipophilic character to distinguish them from the soluble secreted forms. The membrane-bound form in these cases could be an unprocessed primary translate, lipophilic because of the presence of the signal peptide, but some specific lipophilic processing is not quite excluded. Anyhow, there does not seem to be any glyceride-lipoprotein intermediates as in the case of the β-lactamases discussed above. The amino acid sequence of the signal peptides of many of these proteins is known (see Table 1) and shown to lack the specific tetrapeptide at the junction of the signal peptide and mature protein typical to the proper lipoproteins (see Wu and Tokunaga, this volume).

9 Exported Proteins Bound to the Cell Wall and Membrane

A number of bacillar proteins (enzymes) are associated with the cell wall and membrane in an intriguing way. A major fraction of the enzyme is tightly bound to the cytoplasmic membrane, and released from it only by relatively harsh treatments (see below), whereas at least circumstantial evidence suggests that the protein is located on the surface of the cytoplasmic membrane readily accessible from the outside. Examples of such enzymes are the alkaline phosphatases of several bacillar species (see PRIEST 1977, and below), β-glucosidase of *B. licheniformis* (THIRUNAVUKKARASU and PRIEST 1984) and the lipase of *B. subtilis* (KENNEDY and LENNARZ 1979). These enzymes have been studied by several laboratories hoping that some of them might represent novel types of precursors in protein secretion or be secreted by a novel mechanism not found in gram-negative bacteria. The object of the most detailed studies has been the alkaline phosphatases of *B. licheniformis* and *B. subtilis*.

The bulk of the alkaline phosphatase in bacilli (*B. subtilis* and *B. licheniformis* have mainly been studied) is associated with the cell wall. In *B. licheniformis* a few percent of the enzymatic activity is released into the growth medium (SPENCER et al. 1981, 1982). Another portion of the enzyme can be released as soluble protein either by lysing the cells with lysozyme or by converting them to protoplasts without cell breakage (GLYNN et al. 1977; HANSA et al. 1981; SPENCER et al. 1981, 1982; HULETT 1984). The amount thus released and solubilized varied from a few percent up to 80% depending on the growth medium and the phase of growth. It thus seems that at least some of the alkaline phosphatase – a substantial part in the late logarithmic growth phase (GLYNN et al. 1977) – is clearly located on the extracytoplasmic side of the cytoplasmic membrane and associated with it very loosely if at all.

The rest of the enzyme was tightly membrane bound. However, it could be extracted and solubilized almost quantitatively with a high molarity of salt from membrane preparations or from protoplasts (references above). The salt extraction left a variable amount of the enzyme (up to 30% of the membrane-bound enzyme) which could be extracted with mild detergents (SPENCER et al. 1982; KUMAR et al. 1983; GHOSH et al. 1983). By the criteria of enzymatic profile, amino acid composition, and immunological properties the enzymes obtained by the different extraction methods were identical (GLYNN et al. 1977; HANSA et al. 1981). Another enzyme with very similar properties and cellular distribution is the β-glucanase of *B. subtilis* (THIRUNAVUKKARASU and PRIEST 1984).

In *B. subtilis* essentially all of the alkaline phosphatase is in a form tightly bound to the cytoplasmic membrane, but extractable with salt (WOOD and TRISTRAM 1970). A possible indication of the localization of this enzyme outside the cytoplasmic membrane has been obtained when studying a mutant that secretes alkaline phosphatase (HITOTSUYANAGI et al. 1978; YAMANE et al. 1976; YAMANE and MARUO 1978a). Although the mutation has not been genetically characterized, it seems to have affected the protein in such a way that its binding to the cytoplasmic membrane was decreased with consequent release into the medium. This suggests that the original location of alkaline phosphatase also in *B. subtilis* is outside the cytoplasmic membrane.

Histochemical labeling for electron microscopy, based either on enzymatic activity (lead capture labeling) or on specific antibodies with gold label, has shown the presence of alkaline phosphatase on the outer surface of the cytoplasmic membrane and in the matrix of the cell wall in agreement with the above extraction studies (HANSA et al. 1981; SMITH et al. 1981; TINGLU et al. 1984; GHOSH et al. 1983). However, both methods have also consistently revealed alkaline phosphatase on the cytoplasmic side of the cytoplasmic membrane. The gold-labeled immunoglobulins were also found to bind extensively to cytoplasmic material (HANSA et al. 1981). Both labeling techniques indicated that alkaline phosphatase on the inner surface of the cytoplasmic membrane was localized in relatively few clusters. No clusters were seen after extraction with $MgCl_2$ (GHOSH et al. 1983).

The presence of a considerable fraction of alkaline phosphatase inside the cytoplasmic membrane would also be in agreement with the results of studies using vectorial reagents to label the enzyme. Comparison of the specific activity of the label in alkaline phosphatase when protoplasts were labeled before or after disruption of membrane permeability showed a severalfold increase after disruption of protoplasts (SPENCER et al. 1981; GHOSH et al. 1983).

These studies suggest that the alkaline phosphatase of bacilli is located in two cellular compartments. A major fraction of the protein is found in the cytosol, largely bound to the inner surface of the plasma membrane. Another fraction is found outside the plasma membrane, either detached from the membrane (secreted, or releasable by protoplasting) or bound to its outer surface.

Such symmetrical localization of a protein on both sides of a cellular membrane is unique. There are not, however, sufficient data to relate this unusual and interesting situation to the mechanism of secretion or translocation of this protein across the cytoplasmic membrane in more detail. It has been suggested by GHOSH et al. (1983) and TINGLU et al. (1984) that the clustering of alkaline phosphatase to relatively few sites at the inner surface of the cytoplasmic membrane (GHOSH et al. 1983) is suggestive of the presence of specific membrane receptors that could be involved in the process of secretion. There are mutations that either decrease (KUMAR et al. 1983) or increase (YAMANE and MARUO 1978b) the relative amount of secreted alkaline phosphatase. These mutations could have altered the receptor and its binding properties (KUMAR et al. 1983).

However, the available data are not sufficient to propose a detailed mechanism. A major difficulty is that the different forms of alkaline phosphatase have not been characterized beyond their enzymatic and immunological properties. It is thus not possible to establish a clear precursor-product relationship between the forms released by the different extraction procedures, or between the forms in different cellular compartments. Furthermore, as pointed out (HULETT 1984), none of the currently established mechanisms of protein secretion can satisfactorily explain all the findings of the alkaline phosphatase system.

10 Secretion of Foreign Proteins from *B. subtilis*

The high capacity of many species of bacilli to secrete protein has become a focus of biotechnological interest with the hope of utilizing this capacity to produce foreign proteins in secreted form in bacilli. Indeed, as soon as cloning of DNA in bacilli became feasible, such a goal has motivated much of the work performed on the secretion of bacillar proteins. Due to the overall similarity of the mechanism of protein secretion in all organisms studied, bacilli could be expected to be able to secrete at least some proteins of biotechnological interest. The potential advantages of such a technology would be simplified purification, increased yields, and avoidance of the denaturation caused by aggregation seen when proteins are produced intracellularly in high yields. To achieve this goal, many laboratories have explored the straightforward approach of fusing the promoter and signal sequence of a gene of a bacillar secretory protein to the gene or stretch of DNA coding for the foreign protein intended to be secreted.

The first indication of the feasibility of this approach was the production of human insulin by *B. subtilis* (CHANG et al. 1982b). This was achieved by using a plasmid in which a 5′ fragment of the penicillinase gene (*penP*) of *B. licheniformis*, containing the promoter and the signal sequence of *penP*, and a cDNA fragment coding for human preproinsulin were inserted as a tandem fusion. A short deletion was introduced in the junction area to remove an undefined number of codons of the signal peptide of the proinsulin. *B. subtilis* carrying this plasmid synthesized a protein, which reacted with anti-human insulin C-peptide in radioimmune assay. The protein was found both in cellular extracts and in the growth medium. However, no data were given on the amount of the presumed fusion protein secreted, on the molecular structure of the cell-bound and secreted forms, or on the possible presence of lipoprotein intermediate.

The secretion vector so far most extensively characterized is based on the secretory α-amylase of *B. amyloliquefaciens*. This protein is secreted in large amounts by industrial strains (INGLE and ERICKSON 1978). The gene coding for this protein was cloned by PALVA et al. (PALVA 1982) in *B. subtilis* using the multicopy plasmid pUB110. The hybrid plasmid obtained was very stably maintained and the cloned α-amylase gene well expressed in *B. subtilis*. Presumably due to a gene dosage effect, the *B. subtilis* strain carrying this plasmid synthesized and secreted about fivefold more α-amylase than the donor strain of the cloned gene (PALVA 1982). It is thus evident that the promoter and signal sequence of this α-amylase gene are highly compatible and effective also in *B. subtilis*. This hybrid plasmid was then used as a secretion vector by inserting structural genes of foreign proteins immediately downstream from the signal sequence (PALVA et al. 1982). The primary translate of the hybrid gene would thus be composed of the foreign protein fused to the signal sequence of α-amylase.

Three different types of proteins have so far been produced with the aid of this secretion vector. The first set were natural secretory proteins; the prokaryotic, periplasmic β-lactamase of *E. coli* (PALVA et al. 1982) and the eukaryotic, human α_2-interferon (PALVA et al. 1983). The genes were inserted into the secretion vector without their own signal sequence. Introduction of the hybrid vectors resulted in the synthesis and effective (>90%) secretion of both proteins in *B. subtilis* (PALVA et al. 1982, 1983). N-terminal sequencing of the secreted proteins showed that the signal peptides of α-amylase were accurately cleaved from both proteins (SIBAKOV et al. 1984; PALVA et al. 1983). In good agreement with the apparent nonimportance of the residues downstream from the cleavage site in natural secretory proteins, the cleavage always took place after the last alanine residue of the signal peptide, regardless of the ensuing amino acid sequence, which was varied to some extent in the different constructions (PALVA et al. 1983; SIBAKOV et al. 1984).

Another secretion vector very similar to the above one was based on the promoter and the signal sequence of the α-amylase of *B. subtilis* (OHMURA et al. 1984a, b). The synthesis and secretion of *E. coli* β-lactamase (OHMURA et al. 1984a) and mouse interferon (SHIROZA et al. 1985) were achieved also using this construction, although at much lower efficiency than described by PALVA et al. (1982).

The above work demonstrates that fusing a signal peptide of a bacillar exoprotein to foreign proteins can cause their secretion from *B. subtilis*. However, the amount of these proteins recovered in the growth medium was very low, even if they were expressed from a promoter which is very effective in *B. subtilis*. The amounts of β-lactamase and α_2-interferon were at best no more than 10% or 1%, respectively, of the α-amylase produced under comparable conditions (SIBAKOV et al. 1984).

A maximal amount of both β-lactamase (PALVA et al. 1982; SIBAKOV et al. 1984; OHMURA et al. 1984a, b) and α_2-interferon (PALVA et al. 1983) was found in the late logarithmic phase of growth, followed by a decrease, which apparently took place parallel with increasing levels of exoproteases in the later phases of growth. A major role of the exoproteases in causing the low yield is also indicated by data showing that the level and stability of specific β-lactamase mRNA, the rate of its translation and the rate of secretion of the protein were similar to those of α-amylase (ULMANEN et al. 1985). The rate of secretion of β-lactamase was not significantly decreased even in the late stationary phase of growth, when the absolute amount of the enzyme in the culture medium was only 2% of that of α-amylase (ULMANEN et al. 1985).

Efforts to use the above principle and secretion vectors to synthesize foreign nonsecretory proteins in secreted form in *B. subtilis* have been even less successful. In one set of experiments, glycoproteins of the envelopes of Semliki Forest virus (SFV) (E1 and E2 proteins) and of vesicular stomatitis virus (VSV) (G protein) were studied. When synthesized in their natural eukaryotic host cells, the bulk of these proteins are translocated across the appropriate cellular membrane (RER) and reach the extracytoplasmic aqueous milieu. Only a short hydrophobic C-terminal peptide anchors them to the lipid bilayer. If the hydrophobic peptide of the G protein of VSV is deleted, the protein is secreted in soluble form from eukaryotic cells (ROSE and BERGMANN 1982).

To achieve the secretion of these proteins from *B. subtilis*, cDNA copies of these genes were manipulated to delete the segments coding for the hydrophobic anchor peptides. The truncated genes were then fused to the signal sequence of the α-amylase of *B. amyloliquefaciens* in the secretion vector (LUNDSTRÖM 1984, 1985; LUNDSTRÖM et al. 1985). Although these proteins were synthesized in *B. subtilis*, the yield of all proteins was similarly very low, in the range of 0.1% of the amount of α-amylase (SIBAKOV et al. 1984). Furthermore, unlike the natural foreign secretory proteins described above, almost all of these proteins (>90%) remained bound to the membrane fraction. Again, the amount and stability of the mRNA of the one of these proteins studied, the E1 protein of SFV, were not significantly different from that of α-amylase (ULMANEN et al. 1985). Minicells of *B. subtilis*, which secrete no proteins, synthesized approximately equal amounts of E1 protein and α-amylase (ULMANEN et al. 1985). Thus, it seems that again the cause of the low yield was post-translational. All the data would again agree with a very rapid proteolytic degradation of the protein secreted from the cell or translocated across the cytoplasmic membrane. However, it cannot be excluded that these proteins differ in some fundamental although unidentified way from the natural secretory proteins, making their secretion in the bacillar cell very inefficient.

The inability of the bacillar secretory machinery to translocate across the cytoplasmic membrane an exported protein of another organism has recently been demonstrated for outer membrane proteins of *E. coli* (KALLIO et al. 1986). Outer membrane proteins are integral membrane proteins, but due to their location in the cell envelope outside the cytoplasmic membrane they are also exported proteins. Extensive studies in several laboratories have shown that they are inserted into and translocated across the cytoplasmic membrane of *E. coli* very similarly to the soluble periplasmic proteins, the gram-negative analogues of secretory proteins (for recent reviews, see SILHAVY et al. 1983;

Tai, this volume). Using the same secretion vector system which enabled the β-lactamase of *E. coli* to be secreted from *B. subtilis*, both the *ompA* and *ompF* genes of *E. coli* were fused to the signal sequence of α-amylase and introduced into *B. subtilis*. Large amounts of both proteins were synthesized but all of the protein remained in the cell and the signal peptide remained unprocessed (Kallio et al. 1986, and personal communication). There is thus a striking difference between the β-lactamase exported efficiently and the outer membrane proteins seemingly incompatible with the protein export machinery of *B. subtilis*, although both are exported and correctly handled in *E. coli*.

The above experiments prove the feasibility of producing foreign proteins as secretory proteins in bacilli, but they also demonstrate problems encountered in the practical applications. All types of proteins are not likely to be exportable, although so far there are no other clear-cut examples of nonexportable proteins than the gram-negative outer membrane proteins. An important problem is the proteolytic degradation of exported proteins, which at present limits the application of the bacillar secretion vectors to the production of those proteins resistant to bacillar exoproteases. Some of these problems may be solved with the development of secretion vectors based on new promoters, new signal peptides, or new hosts devoid of the exoproteases characteristic of *B. subtilis* (Tsukagoshi et al. 1984b).

11 Conclusions and Perspectives

Recent studies to identify components of the secretion machinery have greatly enhanced the suitability of bacilli as subjects for research on protein secretion. This line of research has both provided evidence for several proteins involved in secretion and established systems employing the powerful combination of genetic and biochemical methods to study secretion in bacilli. It can be expected to contribute in a fundamental way to our understanding of the principles and mechanisms of protein secretion in bacilli as well as in other gram-positive bacteria. Fast progress in this area can be expected, because extension of effective cloning techniques to bacilli has largely removed the earlier obstacle of relatively undeveloped basic genetics in these organisms. Furthermore, the recent establishment, in *E. coli*, of a functional vesicle system for studying protein secretion in vitro (Müller and Blobel 1984; Tai, this volume) anticipates the development of similar methodology in other bacteria that would remove another barrier to the study of the mechanism of protein secretion. These developments would also be expected to provide new and better targeted strategies for the applied research in this area, which has so far suffered from incomplete understanding of the basic mechanisms and physiology involved.

The recent accumulation of data on the primary structure of an increasing number of bacillar exoproteins has been a major advance in the research of protein secretion in bacilli. Comparison at this level of bacillar exoproteins with those of other organisms, both eukaryotic and *E. coli*, has shown that protein secretion of bacilli by and large complies with the general model of secretion based on cleavable signal peptides and (possibly) cotranslational translocation across the cytoplasmic membranes. However, the proteins so far studied have been chosen because of their commercial interest and represent a relatively narrow range of exported bacillar proteins. Cloning and sequencing of other genes coding for fundamentally different types of exported bacillar proteins with different functions and locations in the cell wall can be expected to provide information and model systems to probe for possible novel modes of protein export that would be of general interest and significance. Some good candidates for such alternative export pathways in bacilli have been discussed above, and have also been suggested to exist in gram-negative bacteria (Pugsley and Schwartz 1985).

Another interesting problem for future study will be regulation of the export process and the interaction of exported proteins with the various structures of the cell wall. In general, exported proteins have to traverse this dense barrier. In addition, a number of bacillar exported proteins also remain associated with this complex as its structural components or are actively involved in its synthesis. These interactions indicate finely tuned regulation of the synthesis and export of these proteins by mechanisms that are likely to be directly reflected also in the machinery of protein export. The recent advances in methodology and knowledge of the molecular mechanisms of protein secretion discussed in this review may have provided the means needed to elucidate these interactions. The results can be expected to be of wide significance to basic molecular biology as well as to those aspects of applied bacteriology concerned with interaction of microbes with their environment.

References

Adler L-Å, Arvidson S (1984a) Detection of a mebrane-associated protein on detached membrane ribosomes in *Staphylococcus aureus*. J Gen Microbiol 130:1673–1682

Adler L-Å, Arvidson S (1984b) Immunological cross-reaction between proteins supposed to be involved in protein secretion in *Staphylococcus aureus* and *Bacillus subtilis*. FEMS Microbiol Lett 23:17–20

Altenbern RA (1977) Extreme sensitivity of Staphylococcal enterotoxin B and C production to inhibition by cerulenin. Antimicrobiol Agents Chemother 5:906–908

Aubert E, Klier A, Rapoport G (1985) Cloning and expression in *Escherichia coli* of the regulatory *sacU* gene from *Bacillus subtilis*. J Bacteriol 161:1182–1187

Ayusawa D, Yoneda Y, Yamane K, Maruo B (1975) Pleiotropic phenomena in autolytic enzyme(s) content, flagellation, and simultaneous hyperproduction of extracellular α-amylase and protease in a *Bacillus subtilis* mutant. J Bacteriol 124:459–469

Berkeley RCW, Pepper MP, Caulfield MP, Melling J (1978) The inhibition of *Staphylococcus aureus* enterotoxin production by cerulenin and quinacrine; presumptive evidence for a lipid intermediate/protease release mechanism. FEMS Microbiol Lett 4:103–105

Cancedda R, Schlesinger MJ (1974) Localization of polyribosomes containing alkaline phosphatase nascent polypeptides on membranes of *Escherichia coli*. J Bacteriol 117:290–301

Cantwell BA, McConnell DJ (1983) Molecular cloning and expression of a *Bacillus subtilis* β-glucanase gene in *Escherichia coli*. Gene 23:211–219

Caulfield MP, Berkeley RCW, Pepper EA, Melling J (1979) Export of extracellular levansucrase by *Bacillus subtilis*: inhibition by cerulenin and quinacrine. J Bacteriol 138:345–351

Caulfield MP, Horiuchi S, Tai PC, Davis BD (1984) The 64-kilodalton membrane protein of *Bacillus subtilis* is also present as a multiprotein complex on membrane-free ribosomes. Proc Natl Acad Sci USA 81:7772–7776

Chang CN, Nielsen JBK, Izui K, Blobel G, Lampen JO (1982a) Identification of the signal peptidase cleavage site in *Bacillus licheniformis* prepenicillinase. J Biol Chem 257:4340–4344

Chang S, Gray O, Ho D, Kroyer J, Chang S-Y, McLaughlin J, Mark D (1982b) Expression of eukaryotic genes in *B. subtilis* using signals of *penP*. In: Ganesan AT, Chang S, Hoch JA (eds) Molecular cloning and gene regulation in bacilli. Academic, New York, pp 159–169

Cornelis P, Digneffe C, Willemot K (1982) Cloning and expression of a *Bacillus coagulans* amylase gene in *Escherichia coli*. MGG 186:507–511

Crane LJ, Bettinger GE, Lampen JO (1973) Affinity chromatography purification of penicillinase of *Bacillus licheniformis* 749/C and its use to measure turnover of the cell bound enzyme. Biochem Biophys Res Comm 50:220–227

Debabov VG (1982) The industrial use of Bacilli. In: Dubnau DA (ed) The molecular biology of the Bacilli. Academic, New York, pp 331–370

Fishman Y, Rottem S, Citri N (1978) Evidence linking penicillinase formation and secretion to lipid metabolism in *Bacillus licheniformis*. J Bacteriol 134:434–439

Fishman Y, Rottem S, Citri N (1980) Preferential suppression of normal exoenzyme formation by membrane-modifying agents. J Bacteriol 141:1435–1438

Fouet A, Aubert E, Arnaud M, Klier A, Rapoport G, Le Coq D (1984a) The sucrose system as a model of genetic regulation in *Bacillus subtilis*. In: Ganesan AT, Hoch JA (eds) Genetics and biotechnology of bacilli. Academic, Orlando, pp 113–126

Fouet A, Arnaud M, Klier A, Rapoport G (1984b) Characterization of the precursor form of the exocellular levansucrase from *Bacillus subtilis*. Biochem Biophys Res Comm 119:795–800

Fujii M, Imanaka T, Aiba S (1982) Molecular cloning and expression of penicillinase genes from *Bacillus licheniformis* in the thermophile *Bacillus stearothermophilus*. J Gen Microbiol 128:2997–3000

Fujii M, Takagi M, Imanaka T, Aiba S (1983) Molecular cloning of a thermostable neutral protease gene from *Bacillus stearothermophilus* in a vector plasmid and its expression in *Bacillus stearothermophilus* and *Bacillus subtilis*. J Bacteriol 154:831–837

Fukusaki E, Panbangred W, Shinmyo A, Okada H (1984) The complete nucleotide sequence of the xylanase gene (*xynA*) of *Bacillus pumilus*. FEBS Lett 171:197–201

Gay P, Coq D, Steinmetz M, Ferrari E, Hoch J (1983) Cloning structural gene *sacB*, which codes for exoenzyme levansucrase of *Bacillus subtilis*: expression of the gene in *Escherichia coli*. J Bacteriol 153:1424–1431

Ghosh A, Vallespir S, Ghosh BK (1983) Specificity of subcellular distribution of alkaline phosphatase in *Bacillus licheniformis* 749/C. Can J Microbiol 30:113–125

Glenn AR (1976) Production of extracellular proteins by bacteria. Annu Rev Microbiol 30:41–62

Glynn JA, Schaffel SD, McNicholas JM, Hulett FM (1977) Biochemical localization of the alkaline phosphatase of *Bacillus licheniformis* as a function of culture age. J Bacteriol 129:1010–1019

Halegoua S, Inouye M (1979) Translocation and assembly of outer membrane proteins of *Escherichia coli*. J Mol Biol 130:39–61

Hansa JG, Laporta M, Kuna MA, Reimschuessel R, Hulett FM (1981) A soluble alkaline phosphatase from *Bacillus licheniformis* MC14. Biochim Biophys Acta 657:390–401

Hantke K, Braun V (1973) Covalent binding of lipid to protein. Eur J Biochem 34:284–296

Hitotsuyanagi K, Yamane K, Maruo B (1978) Independent control of the production and localization of alkaline phosphatase from the production of α-amylase and protease in *Bacillus subtilis*. Agric Biol Chem 42:585–592

Honda H, Kudo T, Horikoshi K (1985) Molecular cloning and expression of the xylanase gene of alkalophilic *Bacillus* sp. strain C-125 in *Escherichia coli*. J Bacteriol 161:784–785

Horiuchi S, Marty-Mazars D, Tai PC, Davis BD (1983a) Localization and quantitation of proteins characteristic of the complexed membrane of *Bacillus subtilis*. J Bacteriol 154:1215–1221

Horiuchi S, Tai PC, Davis BD (1983b) A 64-kilodalton membrane protein of *Bacillus subtilis* covered by secreting ribosomes. Proc Natl Acad Sci USA 80:3287–3291

Hulett FM (1984) Cloning and characterization of the *Bacillus licheniformis* gene coding for alkaline phosphatase. J Bacteriol 158:978–982

Ichihara S, Hussain M, Mizushima S (1981) Characterization of new membrane lipoproteins and their precursors of *Escherichia coli*. J Biol Chem 256:3125–3129

Imanaka T (1983) Host-vector systems in thermophilic *Bacilli* and their applications. Trends in Biotechnology 1:139–144

Imanaka T, Tanaka T, Tsunekawa H, Aiba S (1981) Cloning of the genes for penicillinase, *penP* and *penI*, of *Bacillus licheniformis* in some vector plasmids and their expression in *Escherichia coli*, *Bacillus subtilis*, and *Bacillus licheniformis*. J Bacteriol 147:776–786

Imanaka T, Osuihara W, Himeno T, Aiba S (1983) Comparative studies on extracellular penicillinase of the same structural gene, *penP*, expressed in *Bacillus licheniformis* and *Bacillus subtilis*. J Gen Microbiol 129:2621–2628

Ingle MB, Erickson RJ (1978) Bacterial α-amylases. Adv Appl Microbiol 24:257–278

Izui K, Nielsen JBK, Caulfield MP, Lampen JO (1980) Large exopenicillinase, initial extracellular form detected in cultures of *Bacillus licheniformis*. Biochemistry 19:1882–1886

Joyet P, Guérineau M, Heslot H (1984) Cloning of a thermostable α-amylase gene from *Bacillus licheniformis* and its expression in *Escherichia coli* and *Bacillus subtilis*. FEMS Microbiol Lett 21:353–358

Kallio P, Simonen M, Palva I, Sarvas M (1986) Synthesis of OmpA protein of *Escherichia coli* K12 in *Bacillus subtilis*. J Gen Microbiol 132 (in press)

Kennedy MB, Lennarz WJ (1979) Characterization of the extracellular lipase of *Bacillus subtilis* and its relationship to a membrane-lipase found in a mutant strain. J Biol Chem 254:1080–1089

Kreft J, Berger H, Härtlein M, Müller B, Weidinger G, Goebel W (1983) Cloning and expression in *Escherichia coli* and *Bacillus subtilis* of the hemolysin (cereolysin) determinant from *Bacillus cereus*. J Bacteriol 155:681–689

Kroyer J, Chang S (1981) The promoter-proximal region of the *Bacillus licheniformis* penicillinase gene: nucleotide sequence and predicted leader peptide sequence. Gene 15:343–347

Kumar R, Ghosh A, Ghosh BK (1983) Alkaline phosphatase secretion-negative mutant of *Bacillus licheniformis* 749/C. J Bacteriol 154:946–954

Kunst F, Pascal M, Lepesant-Kejzalova J, Lepesant JA, Billault A, Dedonder R (1974) Pleiotropic mutations affecting sporulation conditions and the syntheses of extracellular enzymes in *Bacillus subtilis* 168. Biochimie 56:1481–1489

Lampen JO, Nielsen JBK (1982) Penicillinase and the secretion of proteins by bacilli. In: Ganesan AT, Hoch JA (eds) Molecular cloning and gene regulation in bacilli. Academic, New York, pp 99–109

Lampen JO, Wang W, Mezes PSF, Yang Y-Q (1984) β-Lactamases of bacilli: nature and processing. In: Ganesan AT, Hoch JA (eds) Genetics and biotechnology of bacilli. Academic, New York, pp 129–140

Löfdahl S, Guss B, Uhlen M, Philipson L, Lindberg M (1983) Gene for staphylococcal protein A. Proc Natl Acad Sci USA 80:697–701

Lundström K (1984) Expression of the vesicular stomatitis virus membrane glycoprotein gene in *Bacillus subtilis*. FEMS Microbiol Lett 23:65–70

Lundström K (1985) Expression of viral membrane protein genes in *Bacillus subtilis*. Thesis, University of Helsinki

Lundström K, Palva I, Kääriäinen L, Garoff H, Sarvas M, Pettersson RF (1985) Secretion of Semliki Forest virus membrane glycoprotein E1 from *Bacillus subtilis*. Virus Res 2:69–83

Mäntsälä P, Puntala M (1982) Comparison of levansucrase from *Bacillus subtilis* and from *Bacillus amyloliquefaciens*. FEMS Microbiol Lett 13:395–399

Mäntsälä P, Zalkin H (1979) Membrane-bound and soluble extracellular α-amylase from *Bacillus subtilis*. J Biol Chem 254:8540–8547

Mäntsälä P, Zalkin H (1980) Extracellular and membrane-bound proteases from *Bacillus subtilis*. J Bacteriol 141:493–501

Marty-Mazars D, Horiuchi S, Tai PC, Davis BD (1983) Proteins of ribosome-bearing and free-membrane domains in *Bacillus subtilis*. J Bacteriol 154:1381–1388

Mézes PSF, Wang W, Yeh ECH, Lampen JO (1983) Construction of penP1, *Bacillus licheniformis* 749/C β-lactamase lacking site for lipoprotein modification. J Biol Chem 258:11211–11218

Müller M, Blobel G (1984) In vitro translocation of bacterial proteins across the plasma membrane of *Escherichia coli*. Proc Natl Acad Sci USA 81:7421–7425

Murgier M, Pages C, Lazdunski C, Lazdunski A (1982) Translation control of *ompF*, *ompC* and *lamB* genetic expression during lipid synthesis inhibition of *Escherichia coli*. FEMS Microbiol Lett 13:307–311

Murphy N, McConnell DJ, Cantwell BA (1984) The DNA sequence of the gene and genetic control sites for the excreted *B. subtilis* enzyme β-glucanase. Nucleic Acids Res 12:5355–5367

Nagata Y, Yamaguchi K, Maruo B (1974) Genetic and biochemical studies on cell-bound α-amylase in *Bacillus subtilis* Marburg. J Bacteriol 119:425–430

Neugebauer K, Sprengl R, Schaller H (1981) Penicillinase from *Bacillus licheniformis*: nucleotide sequence of the gene and implications for the biosynthesis of a secretory protein in a gram-positive bacterium. Nucleic Acids Res 9:2577–2588

Nielsen JBK, Lampen JO (1982a) Membrane-bound penicillinases in gram-positive bacteria. J Biol Chem 267:4490–4495

Nielsen JBK, Lampen JO (1982b) Glyceride-cysteine lipoproteins and secretion by gram-positive bacteria. J Bacteriol 152:315–322

Nielsen JBK, Caulfield MP, Lampen JO (1981) Lipoprotein nature of *Bacillus licheniformis* membrane penicillinase. Proc Natl Acad Sci USA 78:3511–3515

Ohmura K, Yamazaki H, Takeichi Y, Nakayama A, Otozai K, Yamane K, Yamasaki M, Tamura G (1983) Nucleotide sequence of the promotor and NH_2-terminal signal peptide region of *Bacillus subtilis* α-amylase gene cloned in pUB110. Biochem Biophys Res Comm 112:678–683

Ohmura K, Nakamura K, Yamazaki H, Shiroza T, Yamane K, Jigami Y, Tanaka H, Yoda K, Yamasaki M, Tamura G (1984a) Length and structural effect of signal peptides derived from *Bacillus subtilis* α-amalyse on secretion of *Escherichia coli* β-lactamase in *B. subtilis* cells. Nucleic Acids Res 12:5307–5319

Ohmura K, Shiroza T, Nakamura K, Nakayama A, Yamane K, Yoda K, Yamasaki M, Tamura G (1984b) A *Bacillus subtilis* secretion vector system derived from the *B. subtilis* α-amylase promoter and signal sequence region, and secretion of *Escherichia coli* β-lactamase by the vector system. J Biochem 95:87–93

Palva I (1982) Molecular cloning of α-amylase gene from *Bacillus amyloliquefaciens* and its expression in *B. subtilis*. Gene 19:81–87

Palva I, Pettersson RF, Kalkkinen N, Lehtovaara P, Sarvas M, Söderlund H, Takkinen K, Kääriäinen L (1981) Nucleotide sequence of the promoter and NH_2-terminal signal peptide region of the α-amylase gene from *Bacillus amyloliquefaciens*. Gene 15:43–51

Palva I, Sarvas M, Lehtovaara P, Sibakov M, Kääriäinen L (1982) Secretion of *Escherichia coli* β-lactamase from *Bacillus subtilis* by the aid of α-amylase signal sequence. Proc Natl Acad Sci USA 79:5582–5586

Palva I, Lehtovaara P, Kääriäinen L, Sibakov M, Cantell K, Schein CH, Kashiwagi K, Weissmann C (1983) Secretion of interferon by *Bacillus subtilis*. Gene 22:229–235

Panbangred W, Kondo T, Negoro S, Shinmyo A, Okada H (1983) Molecular cloning of the genes for xylan degradation of *Bacillus pumilus* and their expression in *Escherichia coli*. MGG 192:335–341

Paton JC, May BK, Elliott W (1980) Cerulenin inhibits production of extracellular proteins but not membrane proteins in *Bacillus amyloliquefaciens*. J Gen Microbiol 118:179–187

Petit-Glatron M-F, Chambert R (1981) Levansucrase of *Bacillus subtilis*: conclusive evidence that its production and export are unrelated to fatty-acid synthesis but modulated by membrane-modifying agents. Eur J Biochem 119:603–611

Petit-Glatron M-F, Chambert R, Steinmetz M (1980) Levansucrase of *Bacillus subtilis*: characterization of a form isolated from phenol-treated cells and activated by Triton X-100. Eur J Biochem 103:189–195

Priest FG (1977) Extracellular enzyme synthesis in the genus *Bacillus*. Bacteriol Rev 41:711–753

Pugsley AP, Schwartz M (1985) Export and secretion of proteins by bacteria. FEMS Microbiol Rev 32:3–38

Randall LL (1983) Translocation of domains of nascent periplasmic proteins across the cytoplasmic membrane is independent of elongation. Cell 33:231–240

Randall LL, Hardy SJS (1977) Synthesis of exported proteins by membrane-bound polysomes from *Escherichia coli*. Eur J Biochem 75:43–53

Rose JK, Bergmann JE (1982) Expression from cloned cDNA of cell-surface secreted forms of the glycoprotein of vesicular stomatitis virus in eukaryotic cells. Cell 30:753–762

Sako T, Tsuchida N (1983) Nucleotide sequence of the staphylokinase gene from *Staphylococcus aureus*. Nucleic Acids Res 11:7679–7693

Shiroza T, Nakazawa K, Tashiro N, Yamane K, Yanagi K, Yamasaki M, Tamura G, Saito H,

Kawade Y, Taniguchi T (1985) Synthesis and secretion of biologically active mouse interferon using a *Bacillus subtilis* α-amylase secretion vector. Gene 34:1–8

Shockman GD, Barrett JF (1983) Structure, function, and assembly of cell walls of gram-positive bacteria. Annu Rev Microbiol 37:501–527

Sibakov M, Palva I (1984) Isolation and the 5′-end nucleotide sequence of *Bacillus licheniformis* α-amylase gene. Eur J Biochem 145:567–572

Sibakov M, Lehtovaara P, Pettersson R, Lundström K, Kalkkinen N, Ulmanen I, Takkinen K, Kääriäinen L, Palva I (1984) Secretion of foreign gene products by the aid of a bacillus secretion vector. In: Ganesan AT, Hoch JA (eds) Genetics and biotechnology of bacilli. Academic, Orlando, pp 153–162

Silhavy TJ, Benson SA, Emr SD (1983) Mechanisms of protein localization. Microbiol Rev 47:313–334

Silver S, Wendt L (1967) Mechanism of action of phenethyl alcohol: breakdown of the cellular permeability barrier. J Bacteriol 93:560–566

Sjödahl J (1977) Structural studies on the four repetitive Fc-binding regions in protein A from *Staphylococcus aureus*. Eur J Biochem 78:471–490

Sleytr UB, Messner P (1983) Crystalline surface layers on bacteria. Annu Rev Microbiol 37:311–339

Sloma A, Gross M (1983) Molecular cloning and nucleotide sequence of the type I β-lactamase gene from *Bacillus cereus*. Nucleic Acids Res 11:4997–5004

Smith WP, Tai P-C, Davis BD (1978) Interaction of secreted nascent chains with surrounding membrane in *Bacillus subtilis*. Proc Natl Acad Sci USA 75:5922–5925

Smith WP, Tai P-C, Davis BD (1979) Extracellular labeling of growing secreted polypeptide chains in *Bacillus subtilis* with diazoiodosulfanilic acid. Biochemistry 18:198–202

Smith WP, Tai C-P, Davis BD (1981) *Bacillus licheniformis* penicillinase: cleavages and attachment of lipid during cotranslational secretion. Proc Natl Acad Sci USA 78:3501–3505

Spencer DB, Chen C-P, Hulett FM (1981) Effect of cobalt on synthesis and activation of *Bacillus licheniformis* alkaline phosphatase. J Bacteriol 145:926–933

Spencer MB, Hansa JG, Stuckmann KV, Hulett FM (1982) Membrane-associated alkaline phosphatase from *Bacillus licheniformis* that requires detergent for solubilization: lactoperoxidase ^{125}I localization and molecular weight determination. J Bacteriol 150:826–834

Steinmetz M, Kunst F, Dedonder R (1976) Mapping of mutations affecting synthesis of exocellular enzymes. MGG 148:281–285

Stephens MA, Ortlepp SA, Ollington JF, McConnell DJ (1984) Nucleotide sequence of the 5′ region of the *Bacillus licheniformis* α-amylase gene: comparison with the *B. amyloliquefaciens* gene. J Bacteriol 158:369–372

Thirunavukkarasu M, Priest FG (1983) Synthesis of α-amylase and β-glucosidase by membrane bound ribosomes from *Bacillus licheniformis*. Biochem Biophys Res Comm 114:677–683

Thirunavukkarasu M, Priest FG (1984) Purification and characterization of an extracellular and a cellular β-glucosidase from *Bacillus licheniformis*. J Gen Microbiol 130:3135–3141

Tinglu G, Ghosh A, Ghosh BK (1984) Subcellular localization of alkaline phosphatase in *Bacillus licheniformis* 749/C by immunoelectron microscopy with colloidal gold. J Bacteriol 159:668–677

Traficante LJ, Lampen JO (1977) Vesicle penicillinase of *Bacillus licheniformis*: existence of periplasmic-releasing factor(s). J Bacteriol 129:184–190

Tsukagoshi N, Ihara H, Yamagata H, Udaka S (1984a) Cloning and expression of a thermophilic α-amylase gene from *Bacillus stearothermophilus* in *Escherichia coli*. MGG 193:58–63

Tsukagoshi N, Tabata R, Takemura T, Yamagata H, Shigezo U (1984b) Molecular cloning of a major cell wall protein gene from protein-producing *Bacillus brevis* 47 and its expression in *Escherichia coli* and *Bacillus subtilis*. J Bacteriol 158:1054–1060

Uhlen M, Guss B, Nilsson B, Gatenbeck S, Philipson L, Lindberg M (1984) Complete sequence of the staphylococcal gene encoding protein A. J Biol Chem 259:1695–1702

Ulmanen I, Lundström K, Lehtovaara P, Sarvas M, Ruohonen M, Palva I (1985) Transcription and translation of foreign genes in *Bacillus subtilis* by the aid of a secretion vector. J Bacteriol 162:176–182

Vasantha N, Thompson LD, Rhodes C, Banner C, Nagle J, Filpula D (1984) Genes for alkaline protease and neutral protease from *Bacillus amyloliquefaciens* contain a large open reading frame between the regions coding for signal sequence and mature protein. J Bacteriol 159:811–819

Watson MEE (1984) Compilation of published signal sequences. Nucleic Acids Res 12:5145–5164

Wells JA, Ferrari E, Henner DJ, Estell DA, Chen EY (1983) Cloning, sequencing and secretion of *Bacillus amyloliquefaciens* subtilisin in *Bacillus subtilis*. Nucleic Acid Res 11:7911–7925

Willemot K, Cornelis P (1983) Growth defects of *Escherichia coli* cells which contain the gene of an α-amylase from *Bacillus coagulans* on a multicopy plasmid. J Gen Microbiol 129:311–319

Wong S-L, Price CW, Goldfarb DS, Doi RH (1984) The subtilisin E gene of *Bacillus subtilis* is transcribed from a δ^{37} promoter in vivo. Proc Natl Acad Sci USA 81:1184–1188

Wood DA, Tristram H (1970) Localization in the cell and extraction of alkaline phosphatase from *Bacillus subtilis*. J Bacteriol 104:1045–1051

Yamane K, Maruo B (1978a) Purification and characterization of extracellular soluble and membrane-bound insoluble alkaline phosphatases possessing phosphodiesterase activities in *Bacillus subtilis*. J Bacteriol 134:100–107

Yamane K, Maruo B (1978b) Alkaline phosphatase possessing alkaline phosphodiesterase activity and other phosphodiesterases in *Bacillus subtilis*. J Bacteriol 134:108–114

Yamane K, Miki T, Saito H, Ikeda Y, Maruo B (1976) Isolation of a mutant secreting extracellular soluble alkaline phosphatase in *Bacillus subtilis*. Agr Biol Chem 40:2181–2185

Yamazaki H, Ohmura K, Nakayama A, Takeichi Y, Otozai K, Yamasaki M, Tamura G, Yamane K (1983) α-Amylase genes (*amy*R2 and *amy*E$^+$) from an α-amylase hyperproducing *Bacillus subtilis* strain: molecular cloning and nucleotide sequences. J Bacteriol 156:327–337

Yang M, Galizzi A, Henner D (1983) Nucleotide sequence of the amylase gene from *Bacillus subtilis*. Nucleic Acids Res 11:237–249

Yang MY, Ferrari E, Henner DJ (1984) Cloning of the neutral protease gene of *Bacillus subtilis* and the use of the cloned gene to create an in vitro-derived deletion mutation. J Bacteriol 160:15–21

Yoneda Y, Maruo B (1975) Mutation in *Bacillus subtilis* causing hyperproduction of α-amylase and proteases and its synergistic effect. J Bacteriol 124:48–54

Yoneda Y, Yamane K, Maruo B (1973) Membrane mutation related to the production of extracellular α-amylase and protease in *Bacillus subtilis*. Biochem Biophys Res Commun 50:765–770

Biogenesis of Lipoproteins in Bacteria

H.C. WU[1] and M. TOKUNAGA[2]

1 The Murein Lipoprotein of *Escherichia coli*

The murein lipoprotein of *Escherichia coli* was first discovered by BRAUN and REHN (1969). Subsequently, the so-called free-form lipoprotein was discovered by INOUYE and his coworkers as one of the major outer membrane proteins in *E. coli* (INOUYE et al. 1972). Among many novel features of the structure and biosynthesis of murein lipoprotein is the presence of a lipoamino acid at its NH_2-terminus, *N*-acyl glyceride-cysteine. The NH_2-terminal glycerylcysteine serves as the attachment site of two ester-linked fatty acids and one amide-linked fatty acid (HANTKE and BRAUN 1973). The ester-linked fatty acids in lipoproteins are similar in composition to those present in bulk phospholipids, including the presence of cyclopropane fatty acids in murein lipoprotein as

[1] Department of Microbiology, Uniformed Services University of the Health Sciences, 4301 Jones Bridge Road, Bethesda, MD 20814-4799, USA
[2] Mitsubishi-Kasei Institute of Life Sciences, Tokyo 194, Japan

Current Topics in Microbiology and Immunology, Vol. 125

well as in phospholipids. Of the amide-linked fatty acids present in lipoprotein, 65% were palmitic acid (HANTKE and BRAUN 1973).

A precursor form of the outer membrane lipoprotein, the prolipoprotein, was identified by Inouye and his colleagues (INOUYE et al. 1977b). The structure of prolipoprotein differs from that of mature lipoprotein in a number of ways. Consequently, a series of post-translational modifications and processing reactions must occur prior to the assembly of the lipoprotein into the outer membrane. These modifications include the proteolytic removal of the signal peptide, the transfer of a glyceryl moiety to the sulfhydryl group of cysteine, *O*-acylation of glycerylcysteine, and the attachment of a fatty acid to the α-NH_2 group of cysteine to generate a fully modified and processed lipoprotein.

1.1 Biosynthesis of the Lipid Moiety in Lipoprotein

In vivo studies have revealed that the biosynthesis of the lipid moiety in lipoprotein proceeds as follows:

Lipid-free protein + phosphatidylglycerol
→ glycerylcysteine-protein + phosphatidic acid
Glycerylcysteine-protein + phospholipid →
glyceride-cysteine-lipoprotein + Lysophospholipid

Evidence for the transfer of a glyceryl moiety from phosphatidylglycerol (PG) to the lipid-free protein acceptor is as follows:

1. The incorporation of [2-^{3}H]glycerol into lipoprotein is dependent upon the incorporation of [2-^{3}H]glycerol into PG (CHATTOPADHYAY and WU 1977; CHATTOPADHYAY et al. 1979a).
2. The glyceryl moiety covalently attached to the sulfhydryl group of cysteine is derived from the nonacylated glycerol of PG (CHATTOPADHYAY and WU 1977).
3. Fusion of labeled phospholipid vesicles with intact cells of *Salmonella typhimurium* results in the transfer of the nonacylated glycerol moiety of PG into lipoprotein. No transfer of [2-^{3}H]glycerol from labeled phosphatidylethanolamine (PE) or cardiolipin (CL) was observed following fusion with phospholipid vesicles containing [2-^{3}H]glycerol-labeled PE or CL, respectively (CHATTOPADHYAY et al. 1979b).
4. The lipoprotein of an *E. coli* mutant altered in lipoprotein structure (*lpp*C77) was found to be deficient in ester-linked fatty acids but contained normal amounts of glycerylcysteine and amide-linked fatty acid (INOUYE et al. 1977a; ROTERING and BRAUN 1977). This observation can best be explained by the sequential transfer of the glyceryl moiety and fatty acyl groups in successive steps of modification reactions.

Evidence for the fatty acyl moieties in phospholipids as the acyl donors to both ester-linked and amide-linked fatty acids in lipoprotein is as follows:

1. Pulse-chase experiments have revealed that both the ester-linked and the amide-linked fatty acid residues in lipoprotein are derived from precursors pres-

ent in a large metabolic pool, such as the acyl moieties in phospholipids (LAI et al. 1980).

2. Incorporation of [^{3}H]palmitate into lipoprotein requires concomitant or prior incorporation of [^{3}H]palmitate into phospholipids (LAI et al. 1980).

3. Lipoprotein synthesized in cerulenin-treated cells is not underacylated, whereas lipopolysaccharide synthesized under the same conditions is deficient in ester-linked fatty acids (WALENGA and OSBORN 1980). This observation, and those listed above, rule out free fatty acid, acyl-CoA, or acyl-ACP as the immediate precursor for the fatty acid moieties in lipoprotein (LAI et al. 1980).

4. HANTKE and BRAUN (1973) have commented on the similarity between the fatty acid composition of the diglyceride moiety in lipoprotein and phospholipids and noted the presence of cyclopropane fatty acid as an ester-linked fatty acid in lipoprotein. Inasmuch as the substrates for cyclopropane fatty acid synthetase are phospholipids but not free fatty acids or CoA esters (THOMAS and LAW 1977), the presence of cyclopropane fatty acids in lipoprotein further supports the transacylation reactions postulated above.

5. Fusion of [^{3}H]palmitate-labeled phospholipid vesicles with intact cells of an *E. coli fad*D mutant (which is defective in acyl-CoA synthetase) has provided further support for the hypothesis that the acyl moieties in phospholipids serve as the donors for both the ester-linked and amide-linked fatty acids in lipoprotein. Among the major glycerophosphatides in *E. coli* (PE, PG, and CL), no specificity was observed regarding the efficacy of any individual phospholipid class as the fatty-acyl donor (LAI and WU 1980), see Note Added in Proof.

While the basic scheme for the biosynthesis of the lipid moiety in lipoprotein is derived from studies of the murein lipoprotein in *E. coli*, the main features of this pathway appear to be shared by other microorganisms including *Mycoplasma capricolum*. In this organism, the absence of phospholipase, the presence of phosphatidylglycerol as the predominant phospholipid species, and the total reliance of the cells on the exogenous supply of fatty acids, have allowed an unambiguous demonstration of a pathway using PG as the glyceryl donor and phospholipid-linked fatty acids as the acyl donors in the modification of membrane proteins with glyceride (DAHL et al. 1983; DAHL and DAHL 1984).

1.2 Modification of Prolipoprotein with Glycerol and Fatty Acids Precedes the Processing of Prolipoprotein by Signal Peptidase

A comparison of the structure of prolipoprotein with that of mature lipoprotein allows the prediction that the prolipoprotein undergoes extensive modification and processing reactions to form the mature lipoprotein. It is not clear, a priori, whether modification precedes processing, or vice versa. It has been established that the modification of prolipoprotein with glycerol and fatty acids not only precedes processing, but is, in fact, a prerequisite for further processing by a signal peptidase. The evidence for this conclusion is summarized below:

1. Globomycin, a cyclic peptide antibiotic, specifically inhibits the processing of prolipoprotein to mature lipoprotein (INUKAI et al. 1978). The prolipoprotein

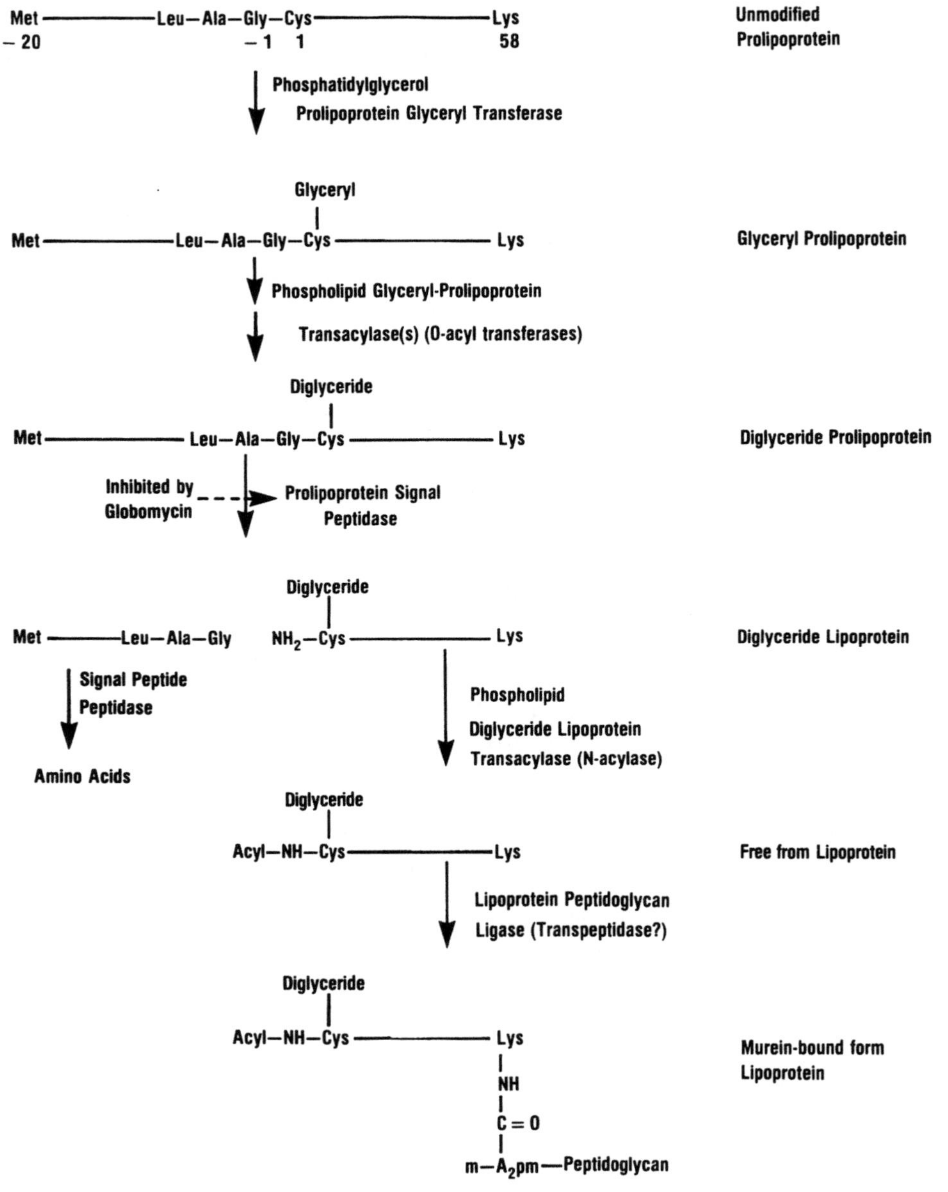

Fig. 1. Pathway for the biosynthesis of murein lipoprotein in *E. coli*. *A_2pm*, meso-diaminopimelic acid

which accumulates in globomycin-treated cells is modified with glyceride (HUSSAIN et al. 1980). Thus, modification of prolipoprotein with glycerol and fatty acids can occur prior to the processing by signal peptidase, at least in globomycin-treated cells.

2. In vivo pulse-chase experiments have revealed the transient appearance of unmodified prolipoprotein, followed by the transient appearance of modified prolipoprotein, and finally the formation of mature lipoprotein (Tokunaga and Wu 1984). The kinetics of labeling of unmodified prolipoprotein, modified prolipoprotein, and mature lipoprotein is consistent with the notion that under unperturbed conditions the modification of prolipoprotein precedes proteolytic cleavage.

3. In vitro studies have shown that unmodified prolipoprotein is not processed by either of the two signal peptidases, SPase I (or M13 procoat protein signal peptidase) and SPase II (prolipoprotein signal peptidase) (Tokunaga et al. 1982b; Dev and Ray 1984). While SPase I does not cleave lipid-modified prolipoprotein, SPase II cleaves lipid-modified prolipoprotein endoproteolytically to form signal peptide and apolipoprotein, which contains glyceride-cysteine at its NH_2-terminus but lacks amide-linked fatty acid (Hussain et al. 1982; Inukai et al. 1984). The amide-linked fatty acid is presumably transferred from phospholipids by a membrane-bound enzyme. The cleavage of lipid-modified prolipoprotein by SPase II is specifically inhibited by globomycin.

Based on both in vivo and in vitro studies, we can formulate the pathway for the biogenesis of lipoprotein in bacteria, as shown in Fig. 1. Inasmuch as the modification of prolipoprotein is the first reaction in this pathway, the specificity of prolipoprotein modification must reside in the amino acid sequence of prolipoprotein including the signal sequence rather than in the primary structure of the mature (processed) protein. As will be discussed later, there is apparently a consensus tetrapeptide sequence of Leu-Ala-Gly-Cys which constitutes the recognition site for the modification and processing enzymes. The presence of this sequence in a precursory protein appears to render this particular protein as a substrate for the lipid modification and subsequent proteolytic processing to become a lipoprotein. Since the cleavage occurs between glycine and glyceride-cysteine, this sequence defines the junction between the signal peptide and the mature protein. Thus, conservation of such a sequence in different precursor proteins which spans the cleavage site strongly suggests the existence of common modification and processing enzymes for a diverse group of lipoproteins which are structurally and functionally distinct (Wu 1985). Had the processing taken place prior to modification, there would have been little sequence homology among different lipoproteins except for the NH_2-terminal cysteine.

2 Lipoprotein as a Generic Term for Fatty Acylated Proteins in Prokaryotic Cells

The studies of murein lipoprotein preceded the discovery of other lipoproteins in bacteria by a decade, partly because of the fact that this major outer membrane lipoprotein is the most abundant protein in *E. coli.* Two key observations led to the realization that the murein lipoprotein is a prototype of fatty acylated and glyceride-modified proteins in both gram-positive and gram-negative bacteria: the discovery of globomycin as a specific inhibitor for lipoprotein processing (Inukai et al. 1978) and the identification of a consensus tetrapeptide in the signal sequences of *E. coli* prolipoprotein and *B. licheniformis* prepenicillinase (Lai et al. 1981b).

The use of globomycin to inhibit the processing of lipid-modified prolipoproteins led to the discovery that multiple, albeit minor, lipoprotein species exist in the cytoplasmic and outer membranes of the *E. coli* cell envelope (Ichihara et al. 1981). These lipoproteins vary in size, number of copies per cell, and subcellular localization, and they show no immunological cross-reactivity with antiserum prepared against purified murein lipoprotein (Ichihara et al. 1981). Using ^{32}P-labeled *lpp* mRNA as the probe, no extensive homology was detected between the *lpp* gene and the genes for the putative new lipoproteins (Nakamura et al. 1979). One can therefore conclude that the enzymes responsible for the post-translational modification reactions must recognize some common structures other than the primary structures of the mature proteins. It is likely that the amino acid sequence located in the vicinity of the modification and cleavage site of these prolipoproteins constitutes such a common structural feature.

It has been shown previously that penicillinase from *B. licheniformis* exists in three forms, the two exoenzymes, exoF and exoS, which are excreted into the culture medium, and the membrane-bound form, which is associated with the cytoplasmic membrane presumably via a hydrophobic segment at the NH_2-terminus (Yamamoto and Lampen 1976; Simons et al. 1978). The structural gene for this enzyme has been cloned, and the DNA sequence determined (Neugebauer et al. 1981). At the end of the hydrophobic segment of the signal sequence of prepenicillinase is the tetrapeptide, -Leu-Ala-Gly-Cys-, which is precisely the same tetrapeptide surrounding the site of modification and processing of prolipoprotein in *E. coli*. Subsequently, it was demonstrated that *B. licheniformis* penicillinase synthesized in *E. coli* is a lipoprotein, which contains an amide-linked fatty acid as well as glyceride-cysteine (Lai et al. 1981b; Nielsen et al. 1981; Smith et al. 1981). Likewise, the membrane-bound penicillinase in *B. licheniformis* is synthesized as a glyceride-modified lipoprotein in *B. subtilis* and the processing of lipid-modified *B. licheniformis* prepenicillinase in *B. subtilis* is inhibited by globomycin (Hayashi and Wu 1983).

The list of lipid-modified proteins is increasing rapidly. Operationally, there are several approaches to ascertain whether a particular protein is a lipoprotein:

1. Labeling of the cells with [2-^{3}H]glycerol or [^{3}H]palmitate to determine whether a particular protein, identified by immunoprecipitation and/or amplification by molecular cloning, is labeled with these precursors. A more definitive assessment of the linkage between glycerol and the protein involves the identification of glycerylcysteine in the acid hydrolysate of the putative lipoprotein. Likewise, the linkage of fatty acids with the protein can be deduced by alkaline and acid hydrolysis of the fatty acid-labeled protein (Lai et al. 1981b).

2. Effect of globomycin on the processing of the precursor protein. Globomycin is a specific inhibitor of SPase II. While the processing of structurally distinct prolipoproteins may be affected by globomycin to different extents, it is nevertheless a simple and reliable indication of the lipoprotein nature of the putative

precursor protein if its processing is inhibited by globomycin (Ichihara et al. 1981). By the same reasoning, the use of a temperature-sensitive *lsp* mutant with a defective SPase II has also allowed the identification of new lipoproteins (Yamagata et al. 1982). Like the effect of globomycin, the processing of distinct prolipoproteins may be affected to different extents in a particular *lsp* mutant at the nonpermissive temperature.

3. Inspection of the signal sequence of the precursor protein. Table 1 lists the signal sequences of known prolipoproteins in bacteria. It is clear that the pres-

Table 1. Signal sequences of lipoprotein precursors in bacteria

		References
Murein Lipoprotein		
E. coli	−20 MetLysAlaThrLysLeuValLeuGlyAla −10 ValIleLeuGlySerThrLeu*LeuAlaGly* +1 *Cys*SerSerAsnAla	Nakamura and Inouye (1979)
S. marcescens	−20 MetAsnArgThrLysLeuValLeuGlyAla −10 ValIleLeuGlySerThrLeu*LeuAlaGly* +1 *Cys*SerSerAsnAla	Nakamura and Inouye (1980)
E. amylovora	−20 MetAsnArgThrLysLeuValLeuGlyAla −10 ValIleLeuGlySerThrLeu*LeuAlaGly* +1 *Cys*SerSerAsnAla	Yamagata et al. (1981)
M. morganii	−20 MetGlyArgSerLysIleValLeuGlyAla −10 ValValLeuAlaSerAlaLeu*LeuAlaGly* +1 *Cys*SerSerAsnAla	Huang et al. (1983)
P. mirabilis	−19 MetLysAlaLysIleValLeuGlyAla −10 ValIleLeuAlaSerGlyLeu*LeuAlaGly* +1 *Cys*SerSerSerAsn	Ching and Inouye (1986)
New lipoproteins in *E. coli*		
Lipoprotein-28	−23 −20 MetLysLeuThrThrHisHisLeuArgThrGlyAlaAla −10 LeuLeuLeuAlaGlyIleLeu*LeuAlaGly* +1 *Cys*AspGlnSerSer	F. Yu et al. (1986)

Table 1 (continued)

		References
Penicillin-binding protein 3 (PBP-3)	−29 MetLysAlaAlaAlaLysThrGlnLys −20 ProLysArgGlnGluGluHisAlaAsnPhe −10 IleSerTrpArgPheAlaLeu*LeuCysGly* +1 *Cys*IleLeuLeuAla	NAKAMURA et al. (1983)
Penicillinase *B. licheniformis*	−26 MetLysLeuTrpPheSer −20 ThrLeuLysLeuLysLysAlaAlaAlaVal −10 LeuLeuPheSerCysValAla*LeuAlaGly* +1 *Cys*AlaAsnAsnGln	NEUGEBAUER et al. (1981)
S. aureus	−16 MetLysLysLeuIlePhe −10 LeuIleValIleAlaLeuVal*LeuSerAla* +1 *Cys*AsnSerAsnSer	McLAUGHLIN et al. (1981)
TraT protein	−20 MetLysMetLysLysLeuMetMetValAla −10 LeuMetSerSerThrLeuAla*LeuSerGly* +1 *Cys*GlyAlaMetSer	OGATA et al. (1982)
Lysis proteins Protein H (pCloDF13)	21 MetLysLysAlaLysAlaIlePheLeuPheIle −10 LeuIleValSerGlyPheLeu*LeuValAla* +1 *Cys*GlnAlaAsnTyr	HAKKAART et al. (1981)
Protein Kil (CelA) (pColE1)	−17 MetArgLysLysPhePheVal −10 GlyIlePheAlaIleAsnLeu*LeuValGly* +1 *Cys*GlnAlaAsnTyr	OKA et al. (1979)
Protein CelB (pColE2)	−19 MetLysLysIleThrGlyIleIleLeu −10 LeuLeuLeuAlaValIleIle*LeuSerAla* +1 *Cys*GlnAlaAsnTyr	COLE et al. (1985)

Table 1 (continued)

		References
Protein Hic (CelC) (pColE3)	−19 MetLysLysIleThrGlyIleIleLeu −10 LeuLeuLeuAlaValIleIle*LeuSerAla* +1 *Cys*GlnAlaAsnTyr	WATSON et al. (1984)
Protein Cal (pColA)	−18 MetLysLysIleIleIleCysVal −10 IleLeuLeuAlaIleMetLeu*LeuAlaAla* +1 *Cys*GlnValAsnAsn	CAVARD et al. (1985)
Pullulanase from *Klebsiella pneumoniae*	−19 MetLeuArgTyrThrCysAsnAlaLeu −10 PheLeuGlySerLeuIleLeu*LeuSerGly* +1 *Cys*AspAsnSerSer	C. CHAPON and O. RAIBAND (manuscript in preparation)
Consensus sequence	−3 +1 Leu*LeuAlaGlyCys*SerSerAsnAla Ala SerAla GlyAlaMetSer Val Val AlaAsnSerAsn Ile AsnGlnLeuGln GlnLeu Tyr AspVal Ile	

ence of Leu-X-Y-Cys at the COOH-terminal end of the signal sequences represents a common feature of the lipoprotein precursors. X and Y are predominantly neutral small amino acids such as alanine, glycine, and serine.

3 Prolipoprotein Modification Enzymes in *E. coli*

Unmodified prolipoprotein accumulates in *E. coli* cells harboring a *mal*E-*lac*Z fused gene upon induction with maltose (ITO et al. 1981). This finding has led to the development of an in vitro system for the modification and processing of prolipoprotein in *E. coli* (TOKUNAGA et al. 1982a). At least four of the enzyme activities involved in the biosynthesis of the mature lipoprotein have been demonstrated in this cell-free system: glyceryltransferase, *O*-acyltransferase(s), prolipoprotein signal peptidase (SPase II), and *N*-acyltransferase. Characterization of the product of the glyceryltransferase reaction in vitro verified that it contained glycerylcysteine (TOKUNAGA et al. 1984a). The optimal pH and reaction temperature for glyceryltransferase were 7.8 and 37° C, respectively. The optimal detergent concentration (Nikkol, non-ionic) for glyceryltransferase was 0.25%. Glyceryltransferase activity in crude cell envelopes is thermostable in the absence of detergent; no loss of activity was seen with pretreatment of cell envelope preparations for 10 min at temperatures up to 65° C. The *N*-acyltransferase appears to be less thermostable than SPase II. Thus, incubation of crude cell envelopes containing glyceride-modified prolipoprotein at 80° C results in the formation of apolipoprotein lacking *N*-acyl fatty acid (HUSSAIN et al. 1982; INUKAI et al. 1984).

Although the glyceryltransferase activity is inactive at pH 5.0, the signal peptidase for prolipoprotein is active at pH 5.0, and both enzymes are active at pH 9.0 (TOKUNAGA et al. 1982a; 1984a). The difference in the pH profiles of glyceryltransferase and SPase II was exploited in an in vitro study to demonstrate that proteolytic processing of prolipoprotein by SPase II requires prior attachment of the glyceride moiety. This conclusion was further supported by studies using purified SPase II. Neither the unmodified prolipoprotein nor the precursor of the maltose-binding protein is cleaved by purified SPase II (DEV and RAY 1984). On the other hand, the glyceride-modified prolipoprotein is efficiently cleaved at the correct site between glycine and modified cysteine by the purified SPase II to yield the processed lipoprotein and the signal peptide in stoichiometric quantities.

The location of the prolipoprotein modification enzymes has not been determined experimentally. It can be inferred, however, from the following observations that they are localized in the cytoplasmic membrane:

1. Mature lipoprotein was detected transiently in the cytoplasmic membrane of *E. coli* cells by pulse-chase experiments (LIN et al. 1980a).
2. Maturation of lipoprotein occured in spheroplasts lacking vestigial outer membrane (KANAZAWA and WU 1979).
3. The lipid-modified prolipoprotein accumulated in the cytoplasmic membrane of globomycin-treated cells (HUSSAIN et al. 1980).
4. SPase II was found to be localized exclusively in the inner membrane (TOKUNAGA et al. 1984a; YAMADA et al. 1984), and the processing of prolipoprotein was shown to require prior modification with lipid (TOKUNAGA et al. 1982a).

4 Processing of Lipid-Modified Prolipoprotein by a Unique Signal Peptidase, the Prolipoprotein Signal Peptidase or Signal Peptidase II

The discovery that globomycin specifically inhibits the processing of prolipoprotein but does not inhibit the processing of the precursors for other major outer membrane proteins provided the first indication that the signal peptidase for prolipoprotein is distinct from that for other precursor proteins.

A signal peptidase (or leader peptidase) has been purified from *E. coli*, based on its activity toward M13 procoat protein (ZWIZINSKI and WICKNER 1980).

It was reported that in addition to M13 procoat protein, the purified signal peptidase I (SPase I) cleaves the precursor forms of the LamB protein, the leucine-specific and isoleucine-valine binding protein, the OmpA protein, the maltose-binding protein, protein III, and lipoprotein (Wolfe et al. 1982; Ito 1982). Biochemical and genetic studies have provided definitive evidence for the existence of at least two distinct signal peptidases (SPase I and SPase II) in *E. coli* (Tokunaga et al. 1982b; Yamagata et al. 1982). More recent work of Ray et al. (p 75–102 of this volume) has suggested the presence of a third signal peptidase in *E. coli* B, which is more active in processing wild-type and mutant maltose-binding protein precursors than SPase I. Purified SPase I does not cleave unmodified or lipid-modified prolipoprotein in vitro (Tokunaga et al. 1982b), and neither the unmodified prolipoprotein nor the precursor of maltose-binding protein is processed by SPase II (Dev and Ray 1984). SPase II activity remained fully active in detergent-solubilized membrane preparations which had been previously treated with antiserum against purified SPase I, whereas the activity toward M13 procoat protein was completely removed by immunoadsorption. *E. coli* cells carrying plasmids with the cloned *lep* gene (the structural gene for SPase I) contain elevated levels of SPase I activity but normal levels of SPase II activity, as compared with the cells with no plasmid (Tokunaga et al. 1982b). An *E. coli* mutant defective in SPase II activity (*lsp*) accumulates a number of lipid-modified precursor proteins at the nonpermissive temperature, but does not accumulate the precursors of other exported proteins (Yamagata et al. 1982). These data taken together clearly indicate that the signal peptidase for prolipoprotein is a distinct processing enzyme which recognizes glyceride-modified cysteine in the lipid-modified precursor proteins (Tokunaga et al. 1984b).

Both SPase I and SPase II have been extensively purified (Wolfe et al. 1982; Dev and Ray 1984). The genes encoding these two enzymes have been cloned and sequenced (Tokunaga et al. 1983; Innis et al. 1984; Yamagata et al. 1983a; Yu et al. 1984b), and their locations in the *E. coli* genetic map have been determined (Regue et al. 1984; Yamagata et al. 1983b). The molecular weights of SPase I and SPase II as determined by sodium dodecyl sulfate polyacrylamide gel electrophoresis (SDS-PAGE) are 37000 and 18000 respectively, and their predicted molecular weights, based on DNA sequences of the *lep* and *lsp* genes, are 35994 and 18144 (Wolfe et al. 1983; Innis et al. 1984; Yu et al. 1984b). SPase II has been shown to be localized exclusively in the cytoplasmic membrane of the *E. coli* cell envelope. The localization of SPase I is less clear. SPase I is found in comparable abundance in the cytoplasmic and outer membranes of wild-type strains of *E. coli*, and the same distribution is found in a strain which overproduces the enzyme four- to tenfold (Zwizinski et al. 1981). In an *E. coli* strain carrying the cloned *lep* gene under thermoinducible regulation, SPase I activity was found to be distributed in both the cytoplasmic (62%–64%) and outer membranes (36%–38%) when the enzyme was not overproduced. When the enzyme level was amplified 75-fold, SPase I activity was increased in both cytoplasmic and outer membranes, and greater than 90% of SPase I activity was localized to the cytoplasmic membrane (Wolfe et al. 1983).

5 Biochemical and Genetic Studies on the Function of SPase I and SPase II In Vivo

SPase I has been characterized as the processing enzyme for M13 procoat protein. In vivo studies with *E. coli* cells that overproduce SPase I have revealed an accelerated conversion of M13 procoat protein to coat protein, especially in mutant M13 phage-infected cells (DATE and WICKNER 1981). Thus, SPase I catalyzes the processing of M13 procoat protein in vivo as well as in vitro (WATTS et al. 1981; OHNO-IWASHITA and WICKNER 1983).

While there has been no report on the isolation of *E. coli* mutants deficient in SPase I activity, negative genetic evidence suggests that SPase I is an essential enzyme in *E. coli* (DATE 1983). Recently, the *lep* gene has been put under the control of P_{ara}, and the cloned gene has been used to replace the chromosomal *lep* gene by genetic recombination (DALBEY and WICKNER 1985). The growth of *E. coli* cells containing a single copy of P_{ara} *lep* appears to be arabinose dependent, suggesting that SPase I activity is essential. Upon removal of arabinose from the culture medium, the growth rate decreases and is eventually accompanied by the accumulation of precursors of exported proteins. Thus, SPase I activity appears to be required for the processing of a number of precursor proteins other than M13 procoat protein. Further biochemical and genetic studies are needed to define the spectrum of the substrates for SPase I in vivo.

The function of SPase II was clarified using globomycin, a specific inhibitor of this enzyme (INUKAI et al. 1978), and by the isolation of a conditionally lethal mutant defective in SPase II activity (YAMAGATA et al. 1982). The accumulation of lipid-modified murein prolipoprotein appears to be correlated with the antibiotic effect of globomycin (LAI et al. 1981 a). Likewise, the accumulation of the precursor forms of minor lipoproteins in globomycin-treated cells and in *lsp* mutant cells at the nonpermissive temperature indicates that SPase II is responsible for the processing of most, if not all, prolipoproteins in *E. coli* (ICHIHARA et al. 1981). This conclusion is further supported by the observation that partially purified SPase II is capable of processing the precursors of new lipoproteins in vitro (YAMADA et al. 1984).

6 Genomic Organizations of the *lep* and *lsp* Genes in *E. coli*

The structural genes for SPase I (*lep*) and SPase II (*lsp*) have been located at 55 min and 0.5 min of the *E. coli* genetic map, respectively (SILVER and WICKNER 1983; REGUE et al. 1984; YAMAGATA et al. 1983b). Unexpectedly, both the *lep* gene and the *lsp* gene are parts of distinct operons.

The *lep* gene is separated from its promoter (DATE and WICKNER 1981; DATE 1983) by an open-reading frame (ORF) encoding a protein with a molecular weight of 66559 (MARCH and INOUYE 1985). The function of this ORF remains obscure.

DNA sequencing of the *lsp* gene has revealed that the initiation codon of the *lsp* gene overlaps the stop codon of the upstream *ileS* gene, the structural

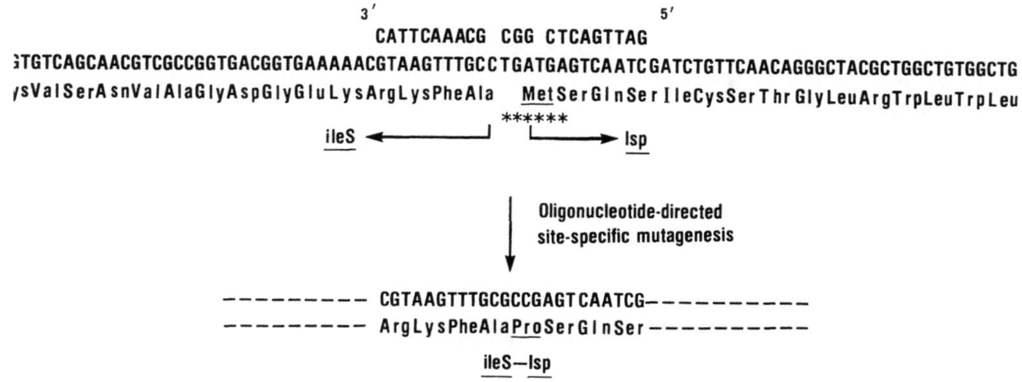

Fig. 2. Fusion of the *ile*S-*lsp* gene. The junction between the *ile*S and the *lsp* genes was changed from CTGATG, which contains the overlapping stop codon of the *ile*S gene and the initiation codon of the *lsp* gene, to GCCG. Consequently, a fusion gene of *ile*S-*lsp* was obtained. The amino acid sequence encoded by the junction region of the hybrid gene was changed from (IleS)-Ala-Stop-Met-Ser-(Lsp) to (IleS)-Ala-Pro-Ser-(Lsp)

gene encoding isoleucyl-tRNA synthetase (INNIS et al. 1984; YU et al. 1984b). This observation suggests the possibilities that the *ile*S and *lsp* genes share a common promoter and the expression of these two genes are translationally coupled. Determination of the nucleotide sequence of the DNA between *ile*S and *rps*T, the structural gene for ribosomal protein S-20, has revealed an opening-reading frame, designated gene *x* (KAMIO et al. 1985). Gene *x* encodes a polypeptide with 312 amino acid residues, and both in vivo and in vitro expression of gene *x* results in the synthesis of a soluble protein with an apparent molecular weight of 35000. The *x*, *ile*S, and *lsp* genes constitute an operon, which is transcribed in the opposite direction to that of the upstream adjacent *rps*T gene. In addition to the promoter at the beginning of this operon, the expression of the *lsp* gene can also be initiated at an internal promoter which is located within the *ile*S gene, approximately 200–400 bp upstream of the *ile*S stop codon (TOKUNAGA et al. 1985). The function of gene *x* remains to be ascertained. The functional significance of the organization of the *lsp* gene as part of an operon encoding proteins involved in protein synthesis and in prolipoprotein processing remains obscure.

The overlap between the stop codon of the *ile*S gene and the initiation codon of the *lsp* gene has allowed the construction of a fusion gene of *ile*S-*lsp*, by altering the junction region from (IleS)-Ala-Stop/Met-Ser-(Lsp) to (IleS)-Ala-Pro-Ser-(Lsp) (TOKUNAGA et al. 1985) (Fig. 2). Cells harboring the fused gene accumulate the hybrid protein in the membrane fraction, whereas the native IleS protein resides in the soluble fraction. These cells also exhibit 25% of SPase II activity as compared with cells harboring the wild-type *lsp* gene, and they show increased resistance to globomycin. These results provide direct evidence that the *ile*S and *lsp* genes are cotranscribed into a single polycistronic mRNA. The data also support the conclusion, based on the inspection of the amino acid sequence of SPase II, that this protein does not contain a cleavable signal peptide at its NH_2-terminus. The localization of the hybrid protein, which contains an extra 111-k Da polypeptide amino-terminal to SPase II, appears to be determined by the topogenic information encoded in the SPase II portion at the COOH-terminus of the hybrid protein. Apparently, the ability of the SPase II domains in the hybrid protein to assemble into the cytoplasmic membrane in a manner that allows the enzymatic function of SPase II has not been severely affected by the addition of a large hydrophilic segment at its NH_2-terminus.

7 Structural Alterations in Prolipoprotein that Affect Modification and Processing Reactions

Mutants altered in the structures of *E. coli* prolipoprotein and *B. licheniformis* prepenicillinase have been isolated or constructed. It is clear from studies of the modification and processing of mutant precursor proteins that the structure and conformation of the precursor proteins affect the modification and processing reactions catalyzed by the enzymes in the lipoprotein pathway or by alternative enzyme(s).

There appear to be at least three structural requirements for the recognition of the precursor proteins by prolipoprotein modification and processing enzymes:

1. The presence of the consensus sequence, Leu-Ala-Gly-Cys or a related sequence, at the modification and processing site
2. The absence of a charged residue in the hydrophobic core of the signal peptide
3. Unique secondary structure flanking the modification and cleavage site

The importance of the consensus sequence Leu-Ala-Gly-Cys has been demonstrated in a number of ways. Replacement of the Cys_{21} with Gly_{21} in the *E. coli* prolipoprotein and removal of Ala_{23}-Leu_{24}-Ala_{25}-Gly_{26}-Cys_{27} in the prepenicillinase result in the synthesis of unmodified and unprocessed precursor forms of lipoprotein and penicillinase respectively (Inouye et al. 1983a, b; Hayashi et al. 1984). In the mutant prepenicillinase (lacking Ala_{23} through Cys_{27}) encoded by the deletion *ΔpenP*2327, the sequence of Leu_{18}-Phe-Ser-Cys-Val_{22} is recognized, albeit poorly, by the prolipoprotein modification enzymes, but the lipid-modified mutant prepenicillinase is not an effective substrate for processing by SPase II. Apparently, the presence of the bulky phenylalanine residue in the tetrapeptide sequence makes this a poor substrate for the prolipoprotein modification and processing enzymes. As shown in Table 1, Leu-Ala-Gly-Cys is the preferred sequence of the consensus tetrapeptide and Leu and Cys are invariable. However, Ser and Ala can replace Ala and Gly, respectively, and the sequences distal to Cys can vary considerably, provided that a β-turn structure immediately following the modification and cleavage site is maintained.

The importance of the consensus tetrapeptide in conferring the specificity of prolipoprotein modification and processing on the polypeptide is further demonstrated by the observation that internalized signal sequences of prolipoprotein are modified and processed in *E. coli* and in *B. subtilis* (Hayashi et al. 1985). An in vitro gene fusion between the erythromycin-resistance gene (*ery*) of *B. subtilis* and the *E. coli lpp* gene encodes, in *E. coli*, three proteins originating from three distinct translation initiation sites within the transcribed mRNA. These three proteins correspond to a hybrid protein containing a 45 amino acid extension preceding the NH_2-terminus of prolipoprotein, a hybrid protein with a 22-amino acid extension preceding the NH_2-terminus of prolipoprotein, and the prolipoprotein, respectively. In *B. subtilis,* the fusion gene encodes two proteins, the hybrid protein with a 45 amino acid extension, and the prolipoprotein. In both *E. coli* and in *B. subtilis,* the precursor forms of the hybrid proteins are lipid modified and are processed to mature lipoprotein in vivo by the globomycin-sensitive SPase II. These results indicate that the presence of an extra 45 amino acid extension containing multiple charged residues beyond the NH_2-terminus of the prolipoprotein signal sequence does not interfere with the modification and processing reactions. Similar results have been obtained with a fusion gene, *lacZ-lpp*, encoding a hybrid β-galactosidase-lipoprotein; the internalized signal sequence with Leu-Ala-Gly-Cys located near the COOH-terminal end of this large hybrid protein is recognized, albeit with greatly reduced efficiency and slower kinetics, by the prolipoprotein modification and processing enzymes (J. Coleman et al., manuscript in preparation).

The importance of unique secondary structure and uninterrupted hydrophobicity in prolipoprotein signal sequence for the proper recognition by prolipoprotein modification enzymes is revealed by the studies of the maturation of prolipoprotein in an *lpp* mutant and its pseudorevertants. The prolipoprotein in an *E. coli* mutant (*lpp*D14) remains unmodified presumably due to the presence of a negatively charged residue in the signal sequence of the prolipoprotein (Lin et al. 1978). Two pseudorevertant alleles of *lpp*D14 have been isolated and characterized (Giam et al. 1984a). In one of the revertants (*lpp*N14), a conversion of Asp_{14} to Asn_{14} restores the modification and processing of the revertant prolipoprotein, albeit with very different kinetics (Tokunaga and Wu 1984). In the other revertant (*lpp*D14I16), a change of Thr_{16} to Ile_{16}, in the presence of the original mutational substitution of Asp_{14}, partially enhances the modification and processing of the revertant prolipoprotein. Secondary structure analysis of the revertant prolipoprotein signal sequences reveals that the specific coil region in residues 14 and 15, the β-sheet structure in residues 16–18, and the β-turn structure at the modification and cleavage site (residue 19–24) of the signal sequence may be important for prolipoprotein modification. Similar structural patterns are also present in the *B. licheniformis* prepenicillinase, the preTraT protein of *E. coli*, and *S. aureus* prepenicillinase, all of which share with the *E. coli* prolipoprotein the same mechanism of modification and processing (Giam et al. 1984a). The importance of a β-turn structure following the modification and cleavage site is also underscored by the alteration in the kinetics of prolipoprotein modification in the *lpp*I23I24 mutant constructed in vitro by oligonucleotide-directed site-specific mutagenesis. A decrease in β-turn structure in the *lpp*$I_{23}I_{24}$ mutant prolipoprotein is accompanied by defective prolipoprotein modification and processing. Modification and processing are further retarded in a triple mutant, *lpp*A20I23I24, by the presence of an additional substitution of Ala_{20} for Gly_{20} (Inouye and Inouye 1985).

Substitution of Gly_9 with aspartic acid (*lpp*D9) results in the accumulation of unmodified prolipoprotein (Pollitt et al. 1985). In contrast, substitution of Gly_9 with arginine (*lpp*R9) does not affect the modification or processing of prolipoprotein (Pollitt et al. 1985). The importance of individual residues in the hydrophobic core of the signal sequence is further illustrated by the defect in prolipoprotein modification and processing in *lpp* mutants constructed by site-directed mutagenesis. Deletion of Gly_{14} (*lpp*∇G14), or Leu_{18} (*lpp*∇L18), affects both modification and processing of prolipoprotein. Substitution of $Ser_{14}Thr_{15}$ with $Ala_{14}Ala_{15}$ (*lpp*A15A16) results in a defective modification while the processing of modified prolipoprotein is not affected. The effect of Leu_{18} deletion is surprising, since a consensus tetrapeptide of Leu_{17}AlaGlyCys is still present in the mutant prolipoprotein. Apparently, a closer juxtaposition of $Ser_{15}Thr_{16}$ to LeuAlaGlyCys in *lpp*∇L18 mutant prolipoprotein somehow affects the modification of prolipoprotein.

Mutations in the mature protein can also affect the kinetics of prolipoprotein modification and processing, as well as the susceptibility of prolipoprotein processing to globomycin inhibition. Thus, both *lpp*C68 and *lpp*C77 mutants are more resistant to globomycin as compared with their parental strains (Lai et al. 1981a; Wu et al. 1982). The kinetics of prolipoprotein modification and processing is slower in *lpp*C68 mutant than that of the wild-type strain (Giam et al. 1984b). The *lpp*C77 mutant lipoprotein appears to be deficient in ester-linked fatty acids (Rotering and Braun 1977; Inouye et al. 1977a), suggestive of a defective *O*-acyl transfer to the mutant prolipoprotein. Thus, the overall conformation of the precursor protein contributes to the recognition of prolipoprotein by modification and processing enzymes.

8 Alternative Processing of Mutant Prepenicillinase and Mutant Prolipoproteins Which Lack the Prolipoprotein Modification Site

Inasmuch as a cysteine residue is required as the acceptor for the glyceryl moiety in a reaction catalyzed by prolipoprotein glyceryltransferase, the first step in the lipoprotein maturation pathway, we can ascertain by the use of mutant precursor proteins lacking the cysteine residue whether *E. coli* cells will process mutationally altered prepenicillinase and prolipoprotein by alternative mechanisms. A number of mutant precursor proteins have been used, and the results can be summarized as follows (Table 2).

1. The ∇*penP*2327 deletion mutant penicillinase is synthesized as an unaltered prepenicillinase; a small fraction (7–9%) of the ∇*penP*2327 mutant prepenicillinase appears to be modified at Cys_{21} to form glyceride-modified prepenicillinase but is not further processed. The mutant prepenicillinase is not secreted into the periplasm, and it is weakly associated with the cell envelope (Hayashi et al. 1984).

Table 2. Alternative processing of unmodified mutant prolipoproteins

Protein	Signal sequence	Modification	Processing	References
Prolipoprotein				
lpp	−20 MetLysAlaThrLysLeuValLeuGlyAla −10 ValIleLeuGlySerThrLeuLeuAlaGly +1 CysSerSerAsn	+	+ ↓ $Gly_{20}Cys_{21}$ [a]	Inouye et al. (1977b)
*lpp*D14	−20 MetLysAlaThrLysLeuValLeuGlyAlaValIleLeu*Asp*SerThrLeuLeuAlaGly +1 CysSerSerAsn	−	−	Lin et al. (1978)
*lpp*N14	−20 MetLysAlaThrLysLeuValLeuGlyAlaValIleLeu*Asn*SerThrLeuLeuAlaGly +1 CysSerSerAsn	+	+ ↓ $Gly_{20}Cys_{21}$ [a]	Giam et al. (1984a)
*lpp*D14I16	−20 MetLysAlaThrLysLeuValLeuGlyAlaValIleLeu*Asp*Ser*Ile*LeuLeuAlaGly +1 CysSerSerAsn	±	± ↓ $Gly_{20}Cys_{21}$ [a]	Giam et al. (1984a)
*lpp*D9	−20 MetLysAlaThrLysLeuValLeu*Asp*AlaValIleLeuGlySerThrLeuLeuAlaGly +1 CysSerSerAsn	−	−	Pollitt et al. (1985)
*lpp*R9	−20 MetLysAlaThrLysLeuValLeu*Arg*AlaValIleLeuGlySerThrLeuLeuAlaGly +1 CysSerSerAsn	+	+ ↓ $Gly_{20}Cys_{21}$ [a]	Pollitt et al. (1985)
*lpp∇*G14	−20 MetLysAlaThrLysLeuValLeuGlyAlaValIleLeu SerThrLeuLeuAlaGly +1 CysSerSerAsn	±	± ↓ $Gly_{20}Cys_{21}$ [a]	Inouye et al. (1984)
*lpp*A15A16	−20 MetLysAlaThrLysLeuValLeuGlyAlaValIleLeuGly*AlaAla*LeuLeuAlaGly +1 CysSerSerAsn	±	+ ↓ $Gly_{20}Cys_{21}$ [a]	Vlasuk et al. (1984)

Table 2 (continued)

Protein	Signal sequence	Modification	Processing	References
lpp∇L18	−20 MetLysAlaThrLysLeuValLeuGlyAlaValIleLeuGlySerThrLeu AlaGly +1 CysSerSerAsn	±	± Gly_{20}↓Cys_{21}[a]	Duffaud et al. (1985)
*lpp*A20	−20 MetLysAlaThrLysLeuValLeuGlyAlaValIleLeuGlySerThrLeuLeuAla*Ala* +1 CysSerSerAsn	+	+ Ala_{20}↓Cys_{21}[a]	Inouye et al. (1983b)
*lpp*I23I24	−20 MetLysAlaThrLysLeuValLeuGlyAlaValIleLeuGlySerThrLeuLeuAlaGly +1 CysSer*IleIle*	+30° C −37° C	+30° C Gly_{20}↓Cys_{21}[a] −37° C	Inouye and Inouye (1985)
*lpp*A20I23I24	−20 MetLysAlaThrLysLeuValLeuGlyAlaValIleLeuGlySerThrLeuLeuAla*Ala* +1 CysSer*IleIle*	−	−	Inouye and Inouye (1985)
lpp-bla	−20 MetLysAlaThrLysLeuValLeuGlyAlaValIleLeuGlySerThrLeuLeuAlaGly +1 CysSerSerAsnAlaLysIleAspGln[GlyIleProGly]-β-lactamase (263 amino acids)	+	Gly_{20}↓Cys_{21}[a]	Ghrayeb and Inouye (1984)
*lpp*G21	−20 MetLysAlaThrLysLeuValLeuGlyAlaValIleLeuGlySerThrLeuLeuAlaGly +1 *Gly*SerSerAsn	−	−	Inouye et al. (1983a)
lpp∇G20	−20 MetLysAlaThrLysLeuValLeuGlyAlaValIleLeuGlySerThrLeuLeuAla +1 CysSerSerAsn	−	−	Inouye et al. (1983b)
*lpp*G21*bla*	−20 MetLysAlaThrLysLeuValLeuGlyAlaValIleLeuGlySerThrLeuLeuAlaGly +1 *Gly*SerSerAsnAlaLysIleAspGln[GlyIleProGly]-β-lactamase (263 amino acids)	−	+ Ala_{25}↓Lys_{26}	Ghrayeb et al. (1985)

lpp∇G20*bla*	−20 MetLysAlaThrLysLeuValLeuGlyAlaValIleLeuGlySerThrLeuLeuAla +1 CysSerSerAsnAlaLysIleAspGln[GlyIleProGly]-β-lactamase (263 amino acids)	−	+ ↓ $Ala_{25}Lys_{26}$	Ghrayeb et al. (1985)
Prepenicillinase				
penP	−26 MetLysLeuTrpPheSer −20 −10 ThrLeuLysLeuLysLysAlaAlaAlaValLeuLeuPheSerCysValAlaLeuAlaGly +1 CysAlaAsnAsnGlnThrAsnAlaSerGlnPro	+	+ ↓ $Gly_{26}Cys_{27}$ [a]	Lai et al. (1981b)
∇*penP*2327	−26 MetLysLeuTrpPheSer −20 ThrLeuLysLeuLysLysAlaAlaAlaValLeuLeuPheSerCysVal +1 AlaAsnAsnGlnThrAsnAlaSerGlnPro	− +(8%, Cys_{21} [a])	− −	Hayashi et al. (1984)
*penP*S27	−26 MetLysLeuTrpPheSer −20 ThrLeuLysLeuLysLysAlaAlaAlaValLeuLeuPheSerCysValAlaLeuAlaGly +1 *Ser*AlaAsnAsnGlnThrAsnAlaSerGlnPro	−	+ ↓ $Ala_{28}Asn_{29}$ ↓ $Ala_{34}Ser_{35}$	Hayashi et al. (1984)
*penP*S27P28	−26 MetLysLeuTrpPheSer −20 ThrLeuLysLeuLysLysAlaAlaAlaValLeuLeuPheSerCysValAlaLeuAlaGly +1 *SerPro*AsnAsnGlnThrAsnAlaSerGlnPro	−	+ ↓ $Ala_{25}Gly_{26}$ ↓ $Ala_{34}Ser_{35}$	Hayashi et al. (1986)
*penP*S27P23′D24′	−26 MetLysLeuTrpPheSer −20 ThrLeuLysLeuLysLysAlaAlaAlaValLeuLeuPheSerCysVal*ProAsp*AlaLeu +1 AlaGly*Ser*AlaAsnAsnGlnThrAsnAlaSerGlnPro	−	−	Hayashi et al. (1986)

[a] Glyceride-modified cysteine

2. The *penP*Ser_{27} penicillinase with a single amino acid substitution of Cys_{27} with Ser_{27} is processed at two distinct sites, between Ala_{28} and Asn_{29} and between and Ala_{34} and Ser_{35} (HAYASHI et al. 1984), and the processed penicillinases are secreted into the periplasm of *E. coli* cells.

3. Two additional mutations in the *penP* gene were derived from *penP*Ser_{27} by site-directed mutagenesis (HAYASHI et al. 1986). One mutant, *penP*Ser_{27}-Pro_{28}, contains a substitution of Ala_{28} with Pro_{28} in addition to the Cys_{27} to Ser_{27} mutation. The presence of proline at position 28 abolishes the processing of prepenicillinase between Ala_{28} and Asn_{29}. Instead, a new site of processing between Ala_{25} and Gly_{26} is unmasked in this mutant along with a second processing site between Ala_{34} and Ser_{25} as found in the *penP*Ser_{27} mutant protein. The other mutant, *penP*$Ser_{27}Pro_{23'}$ $Asp_{24'}$ contains a 6-bp insertion between codons 22 and 23, resulting in the addition of Pro and Asp between Ala_{22} and Val_{23}. The mutant *penP*$Ser_{27}Pro_{23'}$ $Asp_{24'}$ prepenicillinase is not processed. Apparently, the introduction of a charged residue (Asp) and proline upstream of the processing sites drastically alters the structure of the precursor protein so that it is no longer recognized by the alternative processing enzymes.

4. Mutations resulting in alterations of the signal sequence of *E. coli* prolipoprotein have been constructed by oligonucleotide-directed site-specific mutagenesis (POLLITT et al. 1985). Two *lpp* mutations in the prolipoprotein signal sequence, *lpp*ΔG20 and *lpp*G21, resulted in the accumulation of unmodified prolipoprotein. Apparently, there is no alternative pathway of processing of these two mutant prolipoproteins when they are not modified by the lipid modificiation enzymes. On the other hand, when the signal sequences containing these two distinct mutations are joined to the sequence of mature TEM-β-lactamase along with nine amino acids of the mature lipoprotein, the mutated prolipoprotein signal peptides are processed at a site five amino acid residues downstream of the normal modification and processing site between Ala_{25} and Lys_{26} (GHRAYEB et al. 1985). As expected, this alternative processing is not sensitive to globomycin. In contrast, the hybrid protein consisting of the wild-type prolipoprotein signal sequence plus nine amino acids of the mature lipoprotein and the mature β-lactamase, is modified and processed in *E. coli* to form lipo-β-lactamase. In addition, the processing reaction is inhibited by globomycin, which results in the accumulation of lipid-modified prolipo-β-lactamase in the cells (GHRAYEB and INOUYE 1984).

Taken together, the results clearly indicate the complexity in the substrate specificities of precursor-processing enzymes in *E. coli* and other organisms. It is clear that *E. coli*, and presumably other gram-negative and gram-positive bacteria, contains two or more signal peptidases. These include SPase I and SPase II for the processing of nonlipoprotein precursors and prolipoprotein, respectively. The cleavage sites for SPase I and SPase II are distinct, and correspond to the sequences of Ala-X-Ala and Gly$\downarrow$(glyceride-Cys), respectively. Inasmuch as the specificity of SPase II lies in the requirement of glyceride-modified prolipoprotein as its substrate, the specificity of prolipoprotein modification and processing reactions is predicated on the specificity of the modification enzymes, especially the glyceryltransferase.

It is intriguing that mutant prolipoproteins such as *lpp*D14, *lpp*∇ G20, *lpp*G21, and mutant prepenicillinase of ∇*penP*2327 remain unprocessed by SPase I or other processing enzymes in the absence of lipid modification. In contrast, *penP*Ser_{27} prepenicillinase and hybrid prolipo-β-lactamases containing *lpp*∇ G20 or *lpp*G21 are processed at alternative sites following the general pattern of Ala-X-Ala↓ as the processing site. These data strongly suggest that SPase I or equivalent enzymes can process unmodified prolipoproteins provided the following requirements are met: (1) mutational alterations in the precursor protein which preclude competing prolipoprotein modification reactions; (2) the existence of potential cleavage sites for the alternative enzymes (predominantly Ala-X-Ala↓); and (3) structures and conformations in the signal sequence and in the mature protein that are compatible with formation of a functional recognition site on the precursor protein for the alternative processing enzymes. Apparently, potential cleavage sites such as Ala-X-Ala↓ present in mutant prolipoproteins such as ∇*penP*2327, *lpp*G21, and *lpp*∇ G20 precursors, are not sufficiently exposed to allow processing to occur at these sites.

While the signal peptide most likely contains the crucial recognition site for the signal peptidase, structural alterations in the mature protein also affect the cleavage reaction. An inframe deletion in the *lam*B gene removing amino acids 70–200 of the mature λ-receptor protein prevents processing of the deletion mutant precursor by the signal peptidase (Benson and Silhavey 1983). A substitution of Glu by Leu at the second amino acid residue of the mature M13 major coat protein affects the processing of the mutant procoat protein both in vitro and in vivo (Russell and Model 1981). A mutation in the *lpp* gene located near the COOH-terminus of prolipoprotein affects the modification and processing of the mutant prolipoprotein and alters the globomycin susceptibility of the processing reaction by SPase II (Giam et al. 1984b). Thus, the recognition site for the processing enzyme appears to extend beyond the signal sequence. The notion that the tertiary structures of precursor proteins may contribute to the recognition by signal peptidase correlates well with the timing of the processing reaction relative to the extent of completion of the synthesis of the precursor protein. The processing of cotranslationally secreted proteins is a late event and does not occur until the synthesis of the precursor proteins has reached 80% of the full-size polypeptide (Josefsson and Randall 1981). Furthermore, the transmembrane export of nascent chains of secretory proteins does not take place until 80% of the full-size precursor protein is made and the polypeptide is presumably folded into a conformation compatible with the transmembrane translocation (Randall 1983; Randall and Hardy 1984). It is therefore quite reasonable that mutations far removed from the cleavage site may affect the processing of the precursor proteins by signal peptidases by altering the conformations of the precursor proteins.

The identification of alternative processing sites of mutant prolipoproteins at Ala-X, either upstream or downstream of the mutated Gly-Cys sequence, lends credence to the postulate that the cleavage site for signal peptidase(s) is strongly biased toward a sequence of Ala-X-Ala↓ immediately preceding the cleavage site (Perlman and Halvorson 1983). The consensus sequence of small uncharged amino acids at −3 and −1 from the cleavage site has been derived from a statistical analysis of a large number of known signal sequences (VON HEIJNE 1983). In addition to the conserved Ala-X-Ala↓ sequence preceding the cleavage site, the presence of a β-turn secondary structure near the cleavage site as well as a nearly constant six amino acid residue separation between the COOH end of the hydrophobic segment and the cleavage site, may also constitute part of the common structure recognized by the signal peptidase (PERLMAN and HALVORSON 1983).

9 Export of Lipoprotein and Nonlipoproteins Share a Common Pathway or Export Machinery

Protein export is a complex process and requires a number of gene products. In addition to the structural genes for SPase I and SPase II (*lep* and *lsp*, respectively), several genes have been identified which appear to have a general function in protein secretion (*sec*A, *sec*B, secC, and *sec*Y) or protein localization (*prl*A, etc.) (OLIVER and BECKWITH 1981; KUMAMOTO and BECKWITH 1983; FERRO-NOVICK et al. 1984; ITO et al. 1983; EMR et al. 1981; BENSON and SILHAVY 1983). In fact, *prl*A, which is defined as a suppressor for signal sequence mutations in *lam*B and *mal*E, is allelic to *sec*Y; the latter is defined as a *ts* mutation in protein secretion. We can operationally classify precursor proteins into prolipoproteins and nonlipoprotein precursors based on the specificities of their modification and processing enzymes. Since the processing reaction is a late event in protein export, it remains a distinct possibility that the initial events in protein export are common to prolipoproteins and nonlipoprotein precursors alike.

Studies using conditionally lethal mutants defective in protein secretion in general have provided support for the model depicted in Fig. 3. The export of lipoprotein has been found to be affected in both *sec*A (ts) and *sec*Y mutants of *E. coli* but it is not affected in a *sec*B mutant (HAYASHI and WU 1985). The synthesis of lipoprotein is also affected by a mutation in *sec*C (T. Watanabe and H.C. Wu, manuscript in preparation). Thus, the early steps for protein export are common for prolipoproteins and other precursor proteins. These steps presumably involve the recognition of signal peptides by a putative export machinery. The general features of signal peptides in prolipoproteins are much the same as those found in all signal peptides; the major difference lies in the sequence of amino acids in the vicinity of the cleavage site. Divergence of the export pathway for lipoprotein and nonlipoproteins occurs in the cytoplasmic membrane during or immediately after translocation across the membrane. Thus, enzymes involved in post-translational modification or processing will convert precursors with Leu-Ala-Gly-Cys- into lipid-modified prolipoproteins which are subsequently processed by SPase II. These two pathways are not compartmentalized in the cytoplasmic membrane, since unmodifiable mutant prolipo-β-lactamases or prepenicillinases are processed at alternative sites, between Ala and X (HAYASHI et al. 1984; GHRAYEB et al. 1985). These results suggest that precursor proteins exported through the same pathway are accessible to both SPase I and SPase II.

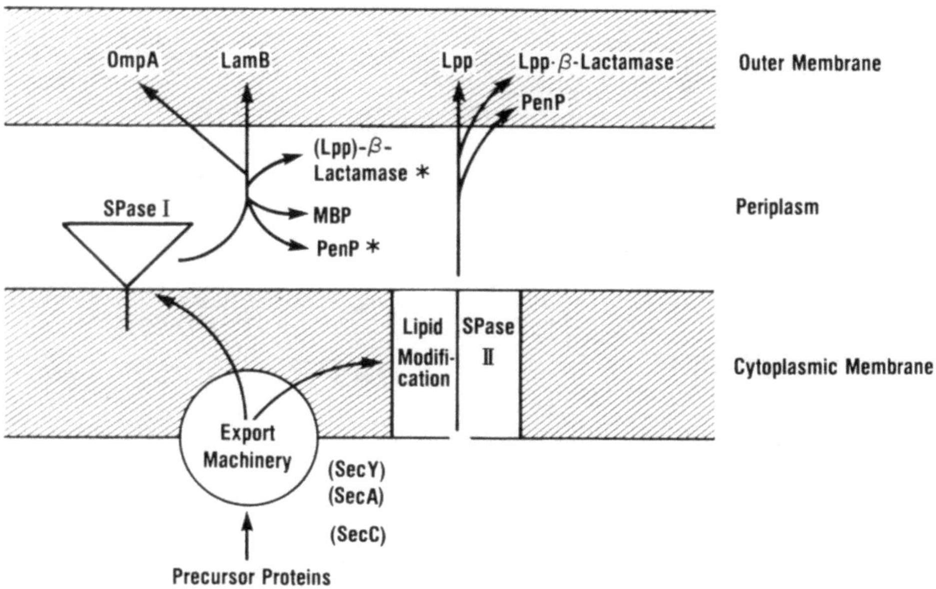

Fig. 3. A common secretory pathway for prolipoprotein and nonlipoprotein precursors in *E. coli*. *PenP** and *(Lpp)-β-lactamase** designate mutant *penP* and *lpp-bla* gene products which are not modifiable, but instead are processed by alternative processing enzymes such as SPase I. (HAYASHI et al. 1984, HAYASHI et al. 1986; GHRAYEB et al. 1985)

10 Assembly of Outer Membrane Lipoprotein

The biogenesis of lipoprotein into the outer membrane can be divided into multiple steps: synthesis of the unmodified prolipoprotein, co- or post-translational translocation of unmodified prolipoprotein through the cytoplasmic membrane, modification and processing of prolipoprotein to form mature lipoprotein, and translocation of mature lipoprotein from the cytoplasmic membrane to the outer membrane (Fig. 4).

1. Synthesis and secretion of unmodified prolipoprotein: In the case of the wild-type prolipoprotein under physiological conditions, the synthesis and secretion of prolipoprotein is efficiently coupled. Alterations in the amino-terminal basic residues of the signal sequence result in the reduction in the synthesis of prolipoprotein as well as the accumulation of unmodified prolipoprotein in the cytoplasm (VLASUK et al. 1983). Apparently, a net negative charge in the NH_2-terminal region of the signal peptide interferes with the synthesis of prolipoprotein as well as the interaction of the signal sequence with the putative export machinery. Translocation of unmodified prolipoprotein into/across the cytoplasmic membrane requires functional SecA and SecY proteins (HAYASHI and WU 1985); post-translational translocation of prolipoprotein is sensitive to CCCP (VLASUK et al. 1983).

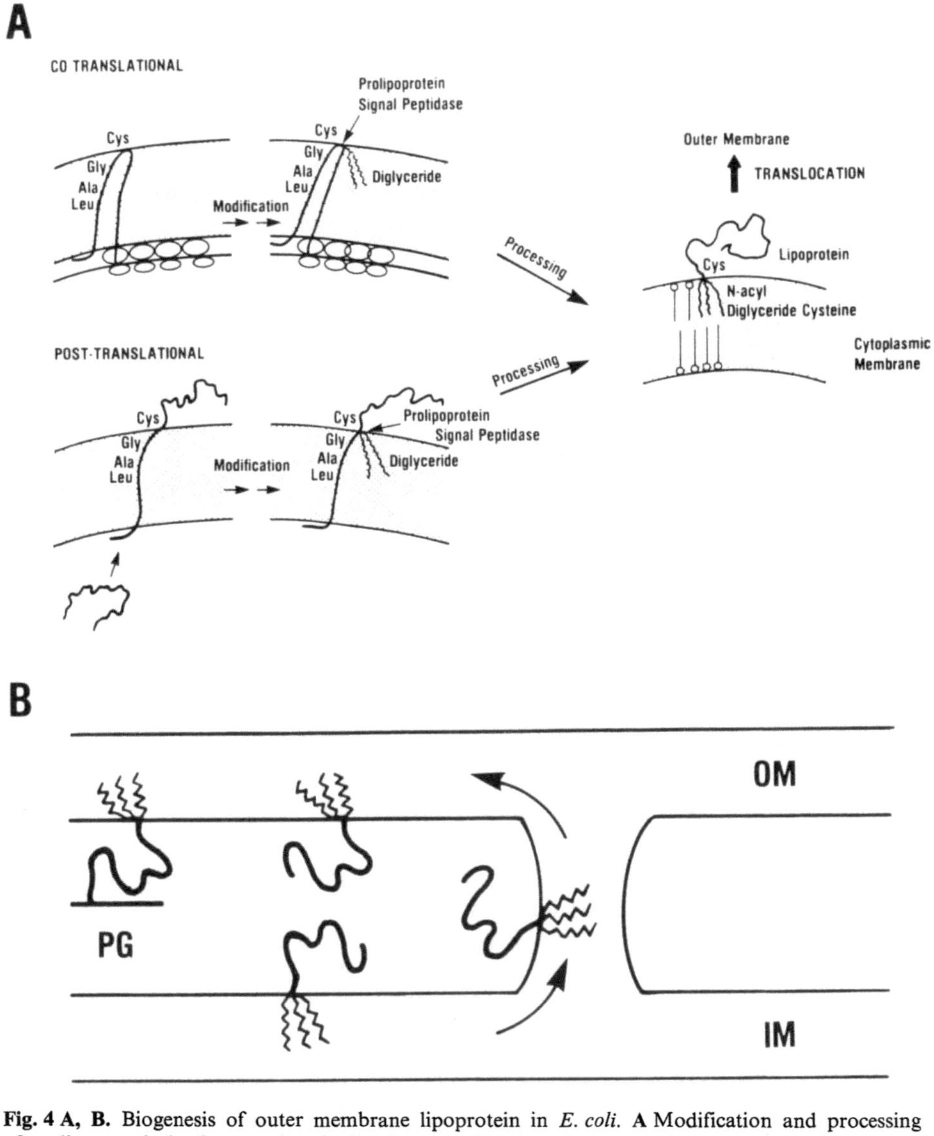

Fig. 4 A, B. Biogenesis of outer membrane lipoprotein in *E. coli*. **A** Modification and processing of prolipoprotein in the cytoplasmic membrane. **B** Intermembrane translocation of mature lipoprotein. *IM*, inner or cytoplasmic membrane; *OM*, outer membrane; *PG*, peptidoglycan

2. Modification and processing of prolipoprotein: Whether modification and processing of prolipoprotein is cotranslational or post-translational remains to be determined (Fig. 4). The small size of murein prolipoprotein, the effect of mutational alteration in the COOH-terminal region of prolipoprotein on the kinetics of prolipoprotein modification and processing (GIAM et al. 1984b), and the demonstration of post-translational modification and processing of prolipoprotein in vitro (TOKUNAGA et al. 1982a), taken together, strongly suggest that modification and processing of prolipoprotein occur post-translationally in vivo.

In vivo studies have suggested that modification and processing take place in the cytoplasmic membrane (LIN et al. 1980a; KANAZAWA and WU 1979). This conclusion is supported by the direct demonstration in vitro of the localization of SPase II in the cytoplasmic membrane (TOKUNAGA et al. 1984a).

3. Translocation and assembly of lipoprotein into the outer membrane. Kinetic studies of the assembly of lipoprotein in *E. coli* have revealed a very rapid and irreversible translocation of lipoprotein from the cytoplasmic membrane to the outer membrane (LIN et al. 1980a). The mechanism of the intermembrane translocation remains totally obscure. It is clear that proteolytic removal of the signal peptide from prolipoprotein is not essential for the translocation and assembly of lipoprotein into the outer membrane. Unmodified mutant prolipoprotein is also rapidly translocated to the outer membrane (LIN et al.1980b). Modified prolipoprotein accumulates in globomycin-treated cells (HUSSAIN et al. 1980). Studies on the localization of lipid-modified prolipoprotein in the cell envelope have provided different answers depending on whether the determination of subcellular localization was based on membrane separation or differential detergent solubilization. It appears that the polypeptide portion of the lipid-modified prolipoprotein is assembled into the outer membrane and covalently attached to the peptidoglycan (INUKAI et al. 1979; ICHIHARA et al. 1982; INUKAI and INOUYE 1983). It is conceivable, however, that the signal peptide and glyceride-modified cysteine in prolipoprotein remains anchored to the cytoplasmic membrane. According to this model, prolipoproteins may exist as transperiplasmic proteins with the signal peptide anchored to the cytoplasmic membrane and the bulk polypeptide interacting with the murein/outer membrane complex. Partition of unmodified or modified prolipoprotein during membrane separation into the outer and cytoplasmic membrane fractions may depend on the presence or absence of the glyceride moiety, as well as on the nature of amino acid alteration in the prolipoprotein. Thus, the interaction of lipid moiety and signal peptide with the cytoplasmic membrane may be strong enough to hold the modified prolipoprotein to the cytoplasmic membrane. In contrast, the association between the signal peptide of unmodified prolipoprotein and the cytoplasmic membrane is weaker, and the unmodified prolipoprotein appears to be associated with both cytoplasmic and outer membranes.

11 The Function of Lipoprotein

The rapidly increasing list of lipid-modified proteins in prokaryotic as well as eukaryotic cells clearly indicates that lipoprotein or fatty acylated protein comprises a diverse group of structurally and functionally distinct proteins. They share a common feature in having a hydrophobic anchor provided by the covalently attached lipid moieties. This common structural feature is derived from a common biosynthetic pathway. Through the NH_2-terminal hydrophobic anchor, these otherwise hydrophilic polypeptides become attached to the cytoplasmic or outer membrane and function accordingly in the topological orientation conferred by both the lipid moiety and the protein itself. At least for lipo-β-lactamase, OmpF-Lpp hybrid protein, and penicillinases, the lipid moiety at the NH_2-termini of these proteins appears to be responsible for their localization and/or anchorage to the outer membrane (GHRAYEB and INOUYE 1984; GHRAYEB et al. 1985; YU et al. 1984a; HAYASHI et al. 1984). Following the reasoning which led to the observation on the importance of being ionized for intracellular retention (DAVIS 1958), one may postulate that covalent modification of proteins with lipids serves to retain otherwise hydrophilic protein to the vicinity of the cellular membrane.

12 Concluding Remarks

Significant progress has been made in the study of the post-translational modification and processing reactions involved in lipoprotein export. Structural features in the signal peptides important for their interaction with the putative export machinery and recognition by the modification and processing enzymes have been delineated. Rapid progress has been made regarding the purification and characterization of signal peptidases from *E. coli*. Biochemical and genetic identification and characterization of prolipoprotein modification enzymes are yet to be accomplished. Among new avenues of future investigation are the in vitro reconstitution of protein secretion including the secretion of lipoproteins (CHEN et al. 1985; MÜLLER and BLOBEL 1984a, b), the characterization of the components involved in protein export process (CHEN and TAI 1985), and studies of the regulation of genes and gene products required for protein export (OLIVER and BECKWITH 1982).

Acknowledgments. This work was supported by National Institutes of Health grant GM-28811 and American Heart Association grant 84-606.

Note Added in Proof

Recent evidence strongly suggests that the amide-linked fatty acid in lipoprotein is derived from the l-acyl moiety in phosphatidyl-ethanolamine (JACKOWSKI S and ROCK CO (1986) J Biol Chem 261, *in press*).

References

Benson SA, Shilhavy TJ (1983) Information within the mature LamB protein necessary for localization to the outer membrane of *E. coli* K12. Cell 32:1325–1335

Braun V, Rehn K (1969) Chemical characterization, spatial distribution and function of a lipoprotein (murein-lipoprotein) of the *E. coli* cell wall. The specific effect of trypsin on the membrane structure. Eur J Biochem 10:426–438

Cavard D, Lloubes R, Morlon J, Chartier M, Lazdunski C (1985) Lysis protein encoded by plasmid ColA-CA31. Gene sequence and export. Mol Gen Genet 199:95–100

Chapon C, Raiband O (1986) Manuscript in preparation

Chattopadhyay PK, Wu HC (1977) Biosynthesis of the covelently linked diglyceride in murein lipoprotein of *Escherichia coli*. Proc Natl Acad Sci USA 74:5318–5322

Chattopadhyay PK, Engle R, Tropp RE, Wu HC (1979a) Biosynthesis of murein lipoprotein in *Escherichia coli*: effects of 3,4-dihydroxybutyl-1-phosphonate. J Bacteriol 138:944–948

Chattopadhyay PK, Lai JS, Wu HC (1979b) Incorporation of phosphatidylglycerol into murein lipoprotein in intact cells of *Salmonella typhimurium* by phospholipid vesicle fusion. J Bacteriol 137:309–312

Chen LL, Tai PC (1985) ATP is essential for protein translocation into *Escherichia coli* membrane vesicles. Proc Natl Acad Sci USA 82:4384–4388

Chen LL, Rhoads D, Tai PC (1985) Alkaline phosphatase and OmpA protein can be translocated posttranslationally into membrane vesicles of *Escherichia coli*. J Bacteriol 161:973–980

Ching G, Inouye M (1986) Expression of the *Proteus mirabilis* lipoprotein gene in *Escherichia coli*. Existence of tandem promoters. J Biol Chem 261:4600–4606

Cole ST, Saint-Joanis B, Pugsley AP (1985) Molecular characterisation of the colicin E2 operon and identification of its products. Mol Gen Genet 198:465–472

Coleman J, Inukai M, Inouye M (1986) Manuscript in preparation

Dahl CE, Dahl JS (1984) Phospholipids as acyl donors to membrane proteins of *Mycoplasma capricolum*. J Biol Chem 259:10771–10776

Dahl CE, Dahl JS, Bloch K (1983) Proteolipid formation in *Mycoplasma capricolum*. Influence of cholesterol on unsaturated fatty acid acylation of membrane. J Biol Chem 258:11814–11818

Dalbey RE, Wickner W (1985) Leader peptidase catalyzes the release of exported proteins from the outer surface of the *Escherichia coli* plasma membrane. J Biol Chem 260:15925–15931

Date T (1983) Demonstration by a novel genetic technique that leader peptidase is an essential enzyme of *Escherichia coli*. J Bacteriol 154:76–83

Date T, Wickner W (1981) Isolation of the *Escherichia coli* leader peptidase gene and effects of leader peptidase overproduction in vivo. Proc Natl Acad Sci USA 78:6106–6110

Davis BD (1958) On the importance of being ionized. Arch Biochem Biophys 78:497–509

Dev IK, Ray PH (1984) Rapid assay and purification of a unique signal peptidase that processes the prolipoprotein from *Escherichia coli* B. J Biol Chem 259:11114–11120

Duffaud GD, Lehnhardt SK, March PE, Inouye M (1985) Structure and function of the signal peptide. In: Knauf PA, Cook IS (eds) Current topics in membrane and transport, vol 24. Academic, London pp 65–101

Emr SD, Hanley-Way S, Silhavy TJ (1981) Suppressor mutations that restore export of a protein with a defective signal sequence. Cell 23:79–88

Ferro-Novick S, Honman M, Beckwith J (1984) The product of gene secC is involved in the synthesis of exported proteins in *E. coli*. Cell 38:211–217

Ghrayeb J, Inouye M (1984) Nine amino acid residues at the NH_2-terminal of lipoprotein are sufficient for its modification, processing, and localization in the outer membrane of *Escherichia coli*. J Biol Chem 259:463–467

Ghrayeb J, Lunn CA, Inouye S, Inouye M (1985) An alternate pathway for the processing of the prolipoprotein signal peptide in *Escherichia coli*. J Biol Chem 260:10961–10965

Giam CZ, Chai T, Hayashi S, Wu HC (1984a) Prolipoprotein modification and processing in *Escherichia coli*. A unique secondary structure in prolipoprotein signal sequence for the recognition by glyceryl transferase. Eur J Biochem 141:331–337

Giam CZ, Hayashi S, Wu HC (1984b) Characterization of a novel lipoprotein mutant in *Escherichia coli*. J Biol Chem 259:5601–5605

Hakkaart MJJ, Veltkamp E, Nijkamp HJJ (1981) Protein H encoded by plasmid Clo DF13 involved in lysis of the bacterial host. I. Localization of the gene and identification and subcellular localization of the gene H product. MGG 183:318–325

Hantke K, Braun V (1973) Covalent binding of lipid to protein. Diglyceride and amide-linked fatty acid at the N-terminal end of the murein-lipoprotein of the *Escherichia coli* outer membrane. Eur J Biochem 34:284–296

Hayashi S, Chang SY, Chang S, Wu HC (1985) Modification and processing of internalized signal sequences of prolipoprotein in *Escherichia coli* and in *Bacilus subtilis*. J Biol Chem 260:5753–5759

Hayashi S, Chang SY, Chang S, Wu HC (1984) Modification and processing of *Bacillus licheniformis* penicillinase in *Escherichia coli*. Fate of mutant penicillinase lacking lipoprotein modification site. J Biol Chem 259:10448–10454

Hayashi S, Chang SY, Chang S, Wu HC (1986) Processing of *Bacillus licheniformis* penicillinases lacking lipoprotein modification site in *Escherichia coli*. J Bacteriol 165:678–681
Hayashi S, Wu HC (1983) Biosynthesis of *Bacillus licheniformis* penicillinase in *Escherichia coli* and in *Bacillus subtilis*. J Bacteriol 156:773–777
Hayashi S, Wu HC (1985) Accumulation of prolipoprotein in *Escherichia coli* mutants defective in protein secretion. J Bacteriol 161:949–954
Huang YX, Ching G, Inouye M (1983) Comparison of lipoprotein gene amoung the Enterobacteriaceae. DNA sequence of *Morganella morganii* lipoprotein gene and its expression in *Escherichia coli*. J Biol Chem 258:8139–8145
Hussain M, Ichihara S, Mizushima S (1980) Accumulation of glyceride-containing precursor of the outer membrane lipoprotein in the cytoplasmic membrane of *Escherichia coli* treated with globomycin. J Biol Chem 255:3707–3712
Hussain M, Ichihara S, Mizushima S (1982) Mechanism of signal peptide cleavage in the biosynthesis of the major lipoprotein of the *Escherichia coli* outer membrane. J Biol Chem 257:5177–5182
Ichihara S, Hussain M, Mizushima S (1981) Characterization of new membrane lipoproteins and their precursors of *Escherichia coli*. J Biol Chem 256:3126–3129
Ichihara S, Hussain M, Mizushima S (1982) Mechanism of export of outer membrane lipoprotein through the cytoplasmic membrane in *Escherichia coli*. Binding of lipoprotein precursors to the peptidoglycan layer. J Biol Chem 257:495–500
Innis MA, Tokunaga M, Williams ME, Loranger JM, Chang SY, Chang S, Wu HC (1984) Nucleotide sequence of the *Escherichia coli* prolipoprotein signal peptidase (lsp) gene. Proc Natl Acad Sci USA 81:3708–3712
Inouye S, Inouye M (1986) Structural requirement at the cleavage site for efficient processing of the lipoprotein secretory precursor of *Escherichia coli*. J Biol Chem 261:(in press)
Inouye M, Shaw J, Shen C (1972) The assembly of a structural lipoprotein in the envelope of *Escherichia coli*. J Biol Chem 247:8154–8159
Inouye S, Lee N, Inouye M, Wu HC, Suzuki H, Nishimura Y, Iketani H, Hirota Y (1977a) Amino acid replacement in a mutant lipoprotein of the *Escherichia coli* outer membrane. J Bacteriol 132:308–313
Inouye S, Wang S, Sekizawa J, Halegoua S, Inouye M (1977b) Amino acid sequence for the peptide extension on the prolipoprotein of the *Escherichia coli* outer membrane. Proc Natl Acad Sci USA 74:1004–1008
Inouye S, Francesechini T, Sato M, Itakura K, Inouye M (1983a) Prolipoprotein signal peptidase of *Escherichia coli* requires a cysteine residue at the cleavage site. EMBO J 2:87–91
Inouye S, Hsu CPS, Itakura K, Inouye M (1983b) Requirement for signal peptide cleavage of *Escherichia coli* prolipoprotein. Science 221:59–61
Inouye S, Vlasuk GP, Hsiung H, Inouye M (1984) Effects of mutations at glycine residue in the hydrophobic region of the *Escherichia coli* prolipoprotein signal peptidase on the secretion across the membrane. J Biol Chem 259:3729–3733
Inukai M, Inouye M (1983) Association of the prolipoprotein accumulated in the presence of globomycin with the outer membrane of *Escherichia coli*. Eur J Biochem 130:27–32
Inukai M, Takeuchi M, Shimizu K, Arai M (1978) Mechanism of action of globomycin. J Antibiot (Tokyo) 31:1203–1205
Inukai M, Ghrayeb J, Nakamura K, Inouye M (1984) Apolipoprotein, an intermediate in the processing of the major lipoprotein of the *Escherichia coli* outer membrane. J Biol Chem 259:757–760
Inukai M, Takeuchi M, Shimizu K, Arai M (1979) Existence of the bound form of prolipoprotein in *Escherichia coli* B cells treated with globomycin. J Bacteriol 140:1098–1101
Ito K (1982) Purification of the precursor form of maltose-binding protein, a periplasmic protein of *Escherichia coli*. J Biol Chem 257:9895–9897
Ito K, Bassford PJ, Beckwith J (1981) Protein localization in *E. coli*: Is there a common step in the secretion of periplasmic and outer-membrane proteins? Cell 24:707–717
Ito K, Wittekind M, Nomura M, Shiba K, Yura T, Miura A, Nashimoto H (1983) A temperature-sensitive mutant of *E. coli* exhibiting slow processing of exported proteins. Cell 32:789–797
Josefsson LG, Randall LL (1981) Different exported proteins in *E. coli* show differences in the temporal mode of processing in vivo. Cell 25:151–157
Kamio Y, Lin CK, Regue M, Wu HC (1985) Characterization of the *ileS-lsp* operon in *Escherichia coli*. Identification of an open reading frame upstream of the *ileS* gene and potential promoter(s) for the *ileS-lsp* operon. J Biol Chem 260:5616–5620

Kanazawa H, Wu HC (1979) Lipoprotein synthesis in *Escherichia coli* spheroplasts: accumulation of lipoprotein in cytoplasmic membrane. J Bacteriol 137:818–823

Kumamoto CA, Beckwith J (1983) Mutations in a new gene, *sec*B, cause defective protein localization in *Escherichia coli.* J Bacteriol 154:253–260

Lai JS, Wu HC (1980) Incorporation of acyl moieties of phospholipids into murein lipoprotein in intact cells of *Escherichia coli* by phospholipid vesicle fusion. J Bacteriol 144:451–453

Lai JS, Philbrick WM, Wu HC (1980) Acyl moieties in phospholipids are the precursors for the fatty acids in lipoprotein of *Escherichia coli.* J Biol Chem 255:5384–5387

Lai JS, Philbrick WM, Hayashi S, Inukai M, Arai M, Hirota Y, Wu HC (1981a) Globomycin sensitivity of *Escherichia coli* and *Salmonella typhimurium*: Effects of mutations affecting structures of murein lipoprotein. J Bacteriol 145:657–660

Lai JS, Sarvas M, Brammar WJ, Neugebauer K, Wu HC (1981b) *Bacillus licheniformis* penicillinase synthesized in *Escherichia coli* contains covalently linked fatty acid and glyceride. Proc Natl Acad Sci USA 78:3506–3510

Lin JJC, Kanazawa H, Ozols J, Wu HC (1978) An *Escherichia coli* mutant with an amino acid alteration within the signal sequence of outer membrane prolipoprotein. Proc Natl Acad Sci USA 75:4891–4895

Lin JJC, Giam CZ, Wu HC (1980a) Assembly of the outer membrane lipoprotein in *Escherichia coli.* J Biol Chem 255:807–811

Lin JJC, Kanazawa H, Wu HC (1980b) Assembly of outer membrane lipoprotein in an *Escherichia coli* mutant with a single amino acid replacement within the signal sequence of prolipoprotein. J Bacteriol 141:550–557

March PE, Inouye M (1985) Characterization of the *lep* operon of *Escherichia coli.* Identification of the promotor and the gene upstream of the signal peptidase I gene. J Biol Chem 260:7206–7213

McLaughlin JR, Murray CL, Rabinowitz JC (1981) Unique features in the ribosome binding site sequence of the gram-positive *Staphylococcus aureus* β-lactamase gene. J Biol Chem 256:11283–11291

Müller M, Blobel G (1984a) In vitro translation of bacterial proteins across the plasma membrane of *Escherichia coli.* Proc Natl Acad Sci USA 81:7421–7425

Müller M, Blobel G (1984b) Protein export in *Escherichia coli* requires a soluble activity. Proc Natl Acad Sci USA 81:7737–7741

Nakamura K, Inouye M (1979) DNA sequence of the gene for the outer membrane lipoprotein of *E. coli*: an extremely AT-rich promotor. Cell 18:1109–1117

Nakamura K, Inouye M (1980) DNA sequence of the *Serratia marcescens* lipoprotein gene. Proc Natl Acad Sci USA 77:1369–1373

Nakamura K, Katz-Wurtzel ET, Pirtle RM, Inouye M (1979) Restriction enzyme cleavage sites surrounding the structural gene for the lipoprotein of the *Escherichia coli* outer membrane. J Bacteriol 138:715–720

Nakamura M, Maruyama IN, Soma M, Kato J, Suzuki H, Hirota Y (1983) On the process of cellular division in *Escherichia coli*: nucleotide sequence of the gene for penicillin-binding protein 3. MGG 191:1–9

Neugebauer K, Sprengel R, Schaller H (1981) Penicillinase from *Bacillus licheniformis*: nucleotide sequence of the gene and implications for the biosynthesis of a secretory protein in a gram-positive bacterium. Nucleic acids Res 9:2577–2588

Nielsen JBK, Caulfield MP, Lampen JO (1981) Lipoprotein nature of *Bacillus licheniformis* membrane penicillinase. Proc Natl Acad Sci USA 78:3511–3515

Ogata RT, Winters C, Levine RP (1982) Nucleotide sequence analysis of the complement resistance gene from plasmid R100. J Bacteriol 151:819–827

Ohno-Iwashita Y, Wickner W (1983) Reconstitution of rapid and asymmetric assembly of M13 procoat protein into liposomes which have bacterial leader peptidase. J Biol Chem 258:1895–1900

Oka A, Nomura N, Morita M, Sugisaki H, Sugimoto K, Takanami M (1979) Nucleotide sequence of small ColE1 derivatives: structure of the regions essential for autonomous replication and ColE1 immunity. Mol Gen Genet 172:151–159

Oliver DB, Beckwith J (1981) *E. coli* mutant pleiotropically defective in the export of secreted proteins. Cell 25:765–772

Oliver DB, Beckwith J (1982) Regulation of a membrane component required for protein secretion in *Escherichia coli.* Cell 30:311–319

Perlman D, Halvorson HO (1983) A putative signal peptidase recognition site and sequence in eukaryotic and prokaryotic signal peptides. J Mol Biol 167:391–409
Pollitt S, Inouye S, Inouye M (1985) Functional analysis of signal peptide mutants. In: Leive L, Bonventre PF, Morello JA, Schlesinger S, Silver SD, Wu HC (eds) Microbiology 1985. American Society for Microbiology, Washington/DC pp 308–311
Randall LL (1983) Translocation of domains of nascent periplasmic proteins across the cytoplasmic membrane is independent of elongation. Cell 33:231–240
Randall LL, Hardy SJS (1984) Export of protein in bacteria. Microbiol Rev 48:290–298
Ray P, Dev I, MacGregor C, Bassford P Jr (1986) Signal peptidases, this volume pp 75–102
Regue M, Remenick J, Tokunaga M, Mackie GA, Wu HC (1984) Mapping of the lipoprotein signal peptidase gene (*lsp*) J Bacteriol 158:632–635
Rotering H, Braun V (1977) Lipid deficiency in a lipoprotein mutant of *Escherichia coli*. FEBS Lett 83:41–44
Russell M, Model P (1981) A mutation downstream from the signal peptidase cleavage site affects cleavage but not membrane insertion of phage coat protein. Proc Natl Aad Sci USA 78:1717–1721
Silver P, Wickner W (1983) Genetic mapping of the *Escherichia coli* leader (signal) peptidase gene (lep): a new approach for determining the map position of a cloned gene. J Bacteriol 154:569–572
Simons K, Sarvas M, Garoff H, Helenius A (1978) Membrane-bound and secreted forms of penicillinase from *Bacillus licheniformis*. J Mol Biol 126:673–690
Smith WP, Tai PC, Davis BD (1981) *Bacillus licheniformis* penicillinase: Cleavages and attachment of lipid during cotranslational secretion. Proc Natl Acad Sci USA 78:3501–3505
Thomas PJ, Law JH (1966) Biosynthesis of cyclopropane compounds. IX Structural and stereochemical requirements for the cyclopropane synthetase substrate. J Biol Chem 241:5013–5018
Tokunaga H, Wu HC (1984) Studies on the modification and processing of prolipoprotein in *Escherichia coli*. Effects of structural alterations in prolipoprotein on its maturation in wild type and *lpp* mutants. J Biol Chem 259:6098–6104
Tokunaga M, Tokunaga H, Wu HC (1982a) Post-translational modification and processing of *Escherichia coli* prolipoprotein in vitro. Proc Natl Acad Sci USA 79:2255–2259
Tokunaga M, Loranger JM, Wolfe PB, Wu HC (1982b) Prolipoprotein signal peptidase in *Escherichia coli* is distinct from the M13 procoat protein signal peptidase. J Biol Chem 257:9922–9925
Tokunaga M, Loranger JM, Wu HC (1983) Isolation and characterization of an *Escherichia coli* clone overproducing prolipoprotein signal peptidase. J Biol Chem 258:12102–12105
Tokunaga M, Loranger JM, Wu HC (1984a) Prolipoprotein modification and processing enzymes in *Escherichia coli*. J Biol Chem 259:3825–3830
Tokunaga M, Loranger JM, Wu HC (1984b) A distinct signal peptidase for prolipoprotein in *Escherichia coli*. J Cell Biochem 24:113–120
Tokunaga M, Loranger JM, Chang SY, Regue M, Chang S, Wu HC (1985) Identification of prolipoprotein signal peptidase and genomic organization of the *lsp* gene in *Escherichia coli*. J Biol Chem 260:5610–5615
Vlasuk GP, Inouye S, Ito H, Itakura K, Inouye M (1983) Effects of the complete removal of basic amino acid residues from the signal peptide on secretion of lipoprotein in *Escherichia coli*. J Biol Chem 258:7141–7148
Vlasuk GP, Inouye S, Inouye M (1984) Effects of replacing serine and threonine residues within signal peptide on the secretion of the major outer membrane lipoprotein of *Escherichia coli*. J Biol Chem 259:6195–6200
Von Heijne G (1983) Patterns of amino acids near signal-sequence cleavage sites. Eur J Biochem 133:17–21
Walenga RW, Osborn MJ (1980) Biosynthesis of lipid A Formation of acyl-deficinet lipopolysaccharides in *Salmonella typhimurium* and *Escherichia coli*. J Biol Chem 255:4257–4263
Watanabe T, Wu HC (1986) Manuscript in preparation
Watson RJ, Lau PCK, Vernet T, Visentin LP (1984) Characterization and nucleotide sequence of a colicin-release gene in the *hic* region of plasmid ColE3-CA38. Gene 29:175–184
Watts C, Silver P, Wickner W (1981) Membrane assembly from purified components II. Assembly of M13 procoat into liposomes reconstituted with purified leader peptidase. Cell 25:347–353
Wolfe PB, Silver P, Wickner W (1982) The isolation of homogeneous leader peptidase from a strain of *Escherichia coli* which overproduces the enzyme. J Biol Chem 257:7898–7902

Wolfe PB, Wickner W, Goodman JM (1983) Sequence of the leader peptidase gene of *Escherichia coli* and the orientation of leader peptidase in the bacterial envelope. J Biol Chem 258:12073–12080
Wu HC (1985) Biogenesis of membrane lipoprotein in Bacteria. In: Leive L, Bonventre PF, Morello JA, Schlesinger S, Silver SD, Wu HC (eds) Microbiology 1985. American Society for Microbiology, Washington DC, pp 303–307
Wu HC, Lai JS, Hayashi S, Giam CZ (1982) Biogenesis of membrane lipoproteins in *Escherichia coli.* Biophys J 37:307–315
Yamada H, Yamagata H, Mizushima S (1984) The major outer membrane lipoprotein and new lipoproteins share a common signal peptidase that exists in the cytoplasmic membrane of *Escherichia coli.* FEBS Lett 166:179–182
Yamagata H, Daishima K, Mizushima S (1983a) Cloning and expression of a gene coding for the prolipoprotein signal peptidase of *Escherichia coli.* FEBS Lett 158:301–304
Yamagata H, Ippolite C, Inukai M, Inouye M (1982) Temperature-sensitive processing of outer membrane lipoprotein in an *Escherichia coli* mutant. J Bacteriol 152:1163–1168
Yamagata H, Nakamura K, Inouye M (1981) Comparison of the lipoprotein gene among the Enterobacteriaceae DNA sequence of *Erwinia amylovora* lipoprotein gene. J Biol Chem 256:2194–2198
Yamagata H, Taguchi N, Daishima K, Mizushima S (1983b) Genetic characerization of a gene for prolipoprotein signal peptidase in *Escherichia coli.* MGG 192:10–14
Yamamoto S, Lampen JO (1976) Membrane penicillinase of *Bacillus licheniformis* 749/C: sequence and possible repeated tetrapeptide structure of the phospholipopeptide region. Proc Natl Acad Sci USA 73:1457–1461
Yu F, Furukawa H, Nakamura K, Mizushima S (1984a) Mechanism of localization of major outer membrane lipoprotein in *Escherichia coli.* Studies with the OmpF-lipoprotein hybrid protein. J Biol Chem 259:6013–6018
Yu F, Inouye S, Inouye M (1986) Lipoprotein-28, a cytoplasmic membrane lipoprotein from *Escherichia coli.* Cloning, DNA sequence, and expression of its gene. J Biol Chem 261:2284–2288
Yu F, Yamada H, Daishima K, Mizushima S (1984b) Nucleotide sequence of the *lspA* gene, the structural gene for lipoprotein signal peptidase of *Escherichia coli.* FEBS Lett 173:264–268
Zwizinski C, Date T, Wickner W (1981) Leader peptidase is found in both the inner and outer membranes of *Escherichia coli.* J Biol Chem 256:3593–3597
Zwizinski C, Wickner W (1980) Purification and characterization of leader (signal) peptidase. J Biol Chem 255:7973–7977

Secretion of Haemolysin by *Escherichia coli*

N. MACKMAN, J.-M. NICAUD, L. GRAY, and I.B. HOLLAND

1 Structure of the *Escherichia coli* Envelope

Escherichia coli secretes very few proteins into the culture medium and it is presumed that the outer membrane constitutes the major barrier to true secretion. The *E. coli* envelope, whose structure has been extensively reviewed (NIKAIDO and NAKAE 1979; OSBORN and WU 1980; HALL and SILHAVY 1981; LUGTENBERG and VAN ALPHEN 1983), is shown in Fig. 1. It is composed of an inner and outer membrane which encloses the peptidoglycan or rigid cell wall. The periplasmic space is also located between the membrane layers. This compartment may contain at least 4% of total cell protein (NOSSAL and HEPPEL 1966) and may have a quite viscous or gel-like structure (HOBOT et al. 1984). The periplasm contains up to 50 distinct polypeptide species (COPELAND et al. 1982), the majority of which are concerned with import mechanisms connecting outer membrane pores (porins) with specific inner membrane permeases. It is

Department of Genetics, University of Leicester, University Road, Leicester LE1 7RH, United Kingdom

Current Topics in Microbiology and Immunology, Vol. 125

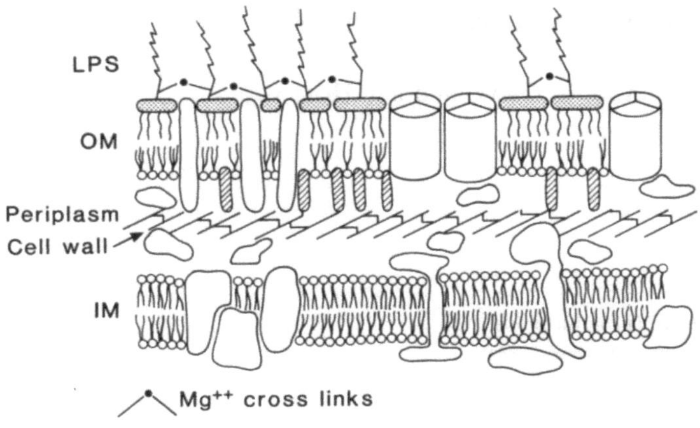

Fig. 1. Structure of the *E. coli* cell envelope

difficult to estimate the precise volume of the periplasmic space under normal growth conditions since it is not structurally defined.

Partial plasmolysis of cells leads to extensive regions of separation between inner and outer membranes and under these conditions numerous connections are observed between the membranes. These structures, identified by BAYER (1979), are still not well characterised and may simply represent regions of biogenesis of outer membrane components within the cytoplasmic membrane. Alternatively, they may constitute more specific export sites to the cell surface. Finally, the plastic nature of both periplasm and inner membrane is highlighted by the observation that bacteria, induced to synthesise large amounts of periplasmic-binding proteins, often show marked expansion of the periplasmic cavity and local distension of the inner membrane, particularly at the cell poles.

The outer membrane of *E. coli* contains two types of lipid molecule distributed asymmetrically; the outer leaflet of the bilayer consists mainly of lipopolysaccharide (LPS) and the inner leaflet consists mainly of phospholipid. The outer membrane appears to be a relatively non-fluid structure, resulting from extensive protein-protein interactions and numerous cross-links formed by magnesium ions between phosphate groups of LPS (NIKAIDO and NAKAE 1979). The resulting surface layer is impermeable to hydrophobic compounds. The bulk of the outer membrane protein is composed of a few species expressed at high levels. These include lipoprotein, OmpA and pore-forming proteins like OmpF and OmpC. Interestingly, unlike the surface proteins of eukaryotic cells, which mostly contain a trans-membrane hydrophobic domain, the proteins of the *E. coli* outer membrane are largely hydrophilic and probably fold back and forth many times across the membrane involving many β-sheet sequences (e.g., CHARBIT et al. 1984). In contrast to the outer membrane, the inner membrane comprises a typical phospholipid bilayer structure which is functionally complex and therefore contains a large variety of protein species.

2 Secretion by Gram-negative Bacteria

Several Gram-negative bacteria secrete a number of hydrolytic enzymes (PUGSLEY and SCHWARTZ 1985) but, so far apparently without exception, when expressed in *E. coli* these polypeptides, which all carry an N-terminal signal sequence, are exported only to the periplasm. Similarly, a number of secreted polypeptides from mammals, such as insulin, when synthesised in *E. coli*, although exported across the inner membrane with concomitant removal of the signal sequence, fail to cross the outer membrane and remain within the periplasm (TALMADGE et al. 1980; TALMADGE and GILBERT 1982). The basis of this outer membrane barrier in *E. coli* to 'true secretion' of polypeptides to the medium is not yet clear. However, it seems most likely that other Gram-negative bacteria which secrete a number of proteins, initially synthesised with an N-terminal signal sequence, contain specific additional factors in the envelope which promote the translocation of individual or 'families' of such polypeptides across the outer membrane from the periplasm. Recently, mutants of *Aeromonas hydrophila* and *Erwinia chrysanthemi* have been isolated which appear to be defective in this last step in secretion (HOWARD and BUCKLEY 1983; ANDRO et al. 1984).

2.1 Possible Secretion Strategies in *E. coli*

Most secreted proteins including *E. coli* periplasmic proteins are water soluble, often quite polar molecules and there are a number of possible strategies which might be used for the transfer of proteins into the culture medium. One possible route for secretion might depend initially upon *secA*, a component of the major export pathway for envelope proteins (OLIVER and BECKWITH 1981), which could transfer the protein via interaction with an N-terminal signal sequence into the periplasmic space. Specific (topogenic) sequences might then be required to recognize outer membrane export 'sites'. In principle, there are apparently precedents for this mechanism in the assembly of surface pili (DODD et al. 1984).

Less-clear possibilities for secretion mechanisms might include direct export from the cytoplasm through 'Bayer junctions' modified by 'secretion-specific' polypeptides which then direct the protein into the medium. Alternatively, some form of vesicular transport mechanism, analogous to that present in eukaryotes (CARO and PALADE 1964) involving topogenic sequences within the polypeptide which recognise 'coated', budding vesicles, may translocate specific polypeptides to the outer membrane followed by release of the lumen content to the medium. As indicated above such a mechanism may be limited by the physical constraints of the layers surrounding the periplasm, although given the fluidity and plasticity of the inner membrane this mechanism cannot be ruled out.

Finally, proteins may be exported by processes which involve partial disruption of the integrity of the surface envelope leading to the gradual but nevertheless preferential release of certain polypeptides. This mechanism, which facilitates the release of colicin molecules, usually results in eventual lysis of the cells (CAVARD et al. 1981; PUGSLEY and SCHWARTZ 1984). An example of this mechanism is discussed in detail in this volume for the release of cloacin DF13 from *E. coli*.

3 Epidemiology and Mode of Action of *E. coli* Haemolysin

At least two forms of haemolysin α and β have previously been reported to be produced by *E. coli* (SMITH 1963), an extracellular and a cell-bound form respectively. For a review see CAVALIERI et al. (1984). Initial results using antiserum indicated that the two haemolysins might be different (SMITH 1963), raising the possibility of two different release mechanisms, but this has not been further substantiated. Moreover, α and β forms are produced simultaneously by *E. coli* during the growth cycle (SNYDER and KOCH 1966) and four independently cloned haemolysin determinants have all proved to be essentially identical in structure. Thus, we propose to simplify the terminology and refer only to '*E. coli* haemolysin'.

Haemolytic strains of *E. coli* characteristically cause haemolysis (clear zone of lysis) around bacterial colonies grown overnight on agar plates containing, for example, sheep blood. In addition, the toxin can be assayed quantitatively by the release of haemoglobin from erythrocytes (MACKMAN and HOLLAND 1984a). Some previous reports also indicate that haemolysin may have cytotoxic effects on a variety of cell types including mouse fibroblast (3T3) cells (CAVALIERI and SNYDER 1982a), human peripheral leucocytes (CAVALIERI and SNYDER 1982b), human neutrophils (R. Welch personal communication) and human blood granulocytes (GADEBERG and ORSKOV 1984). However, the haemolysin used in most of these studies was poorly characterized and as indicated below was likely to have contained proteolytically degraded haemolysin. We now know that the haemolysin is extremely unstable (NICAUD et al. 1985a). In contrast, with the recent availability of highly purified and intact preparations of active haemolysin (see below) considerable progress can be expected in understanding its mode of action. Indeed, some initial studies indicate that *E. coli* haemolysin has a colicin-like action upon erythrocytes with apparently a single bound haemolysin molecule producing a small pore in the surface which may be sufficient to kill the cell (S. Bhakdi, personal communication; see also JORGENSEN et al. 1980).

Haemolysin has been implicated as an important factor in the virulence of pathogenic *E. coli* strains (WELCH et al. 1981). Thus, a significantly high proportion of strains isolated from human urinary tract infections are haemolytic, especially those associated with pyelonephritis (CAVALIERI et al. 1984). Interestingly, most haemolysin determinants carried by *E. coli* isolated from humans have been shown to reside on the chromosome (MÜLLER et al. 1983). In contrast, Hly determinants expressed by strains isolated from animals, although very similar in structure, are almost invariably carried on large transmissible plasmids (MÜLLER et al. 1983). Recent evidence (ZABALA et al. 1984) indicates that the Hly determinant may constitute a transposon-like element. This may explain the alternative locations although not their functional significance. Finally, the widespread occurrence of the Hly determinant is reflected by the presence of circulating antibody to the *E. coli* haemolysin in human populations (S. Bhakdi, personal communication).

4 Production of Haemolysin

Before discussing the possible mechanism of secretion of haemolysin, it is necessary to consider the production and nature of the molecule itself since this has been the subject of some controversy. In the laboratory the highest levels of haemolytic activity are obtained in late exponentially growing cultures of *E. coli* although release is not due to cell lysis (Springer and Goebel 1980; Mackman and Holland 1984a). Appearance in the medium is followed by rapid loss of haemolytic activity as the cells enter stationary phase. The haemolysin is synthesized as a constant fraction of cell protein under steady state conditions when either wild type or laboratory strains carry an Hly determinant (Nicaud et al. 1985a). However, as cultures approach the last phase of growth, synthesis of haemolysin is switched off and the extreme lability of the toxin quickly results in reduction of the extracellular activity to a low level. This switch-off of haemolysin synthesis appears to operate at the level of transcription as revealed by analysis of the expression of an *hlyC-lacZ* hybrid gene (Nicaud et al. 1985a).

Several earlier reports (Jorgensen et al. 1976) indicated that the synthesis of haemolysin was dependent upon a number of possible 'factors' present in rich medium. It was reported that 10- to 100-fold higher levels of haemolytic activity were produced from *E. coli* strains grown in rich medium compared with minimal medium (Snyder and Koch 1966; Jorgensen et al. 1976). Our studies have shown that the addition of 10 m*M* calcium ions to the medium leads to higher levels of haemolysin but this is due to increased stability of the toxin rather than an effect on synthesis (Nicaud et al. 1985a). Moreover, detailed studies, involving the analysis of *hlyC-lacZ* gene fusions and the parallel measurement of both the haemolysin protein and the haemolytic activity in the culture medium, have clearly demonstrated that medium-dependent variations in haemolysin production can now be entirely ascribed to greater or lesser lability of the toxin (Nicaud et al. 1985a). Haemolysin synthesis is also not subject to regulation by catabolite repression (Juarez et al. 1984; Nicaud et al. 1985a). On the other hand it has been reported that in rare cases expression of the Hly determinant is subject to regulation by ferric ions (H. Grunig, personal communication). This may reflect the requirement of some pathogenic *E. coli* strains for an increase in the concentration of soluble ferric ions at the site of infection by the release of haemoglobin from erythrocytes in order to allow growth.

4.1 *E. coli* Haemolysin Is Secreted to the Medium as a 107-kD Polypeptide

Many early attempts to purify and characterize *E. coli* haemolysin were frustrated by its extreme lability and its apparent tendency to aggregate or to bind

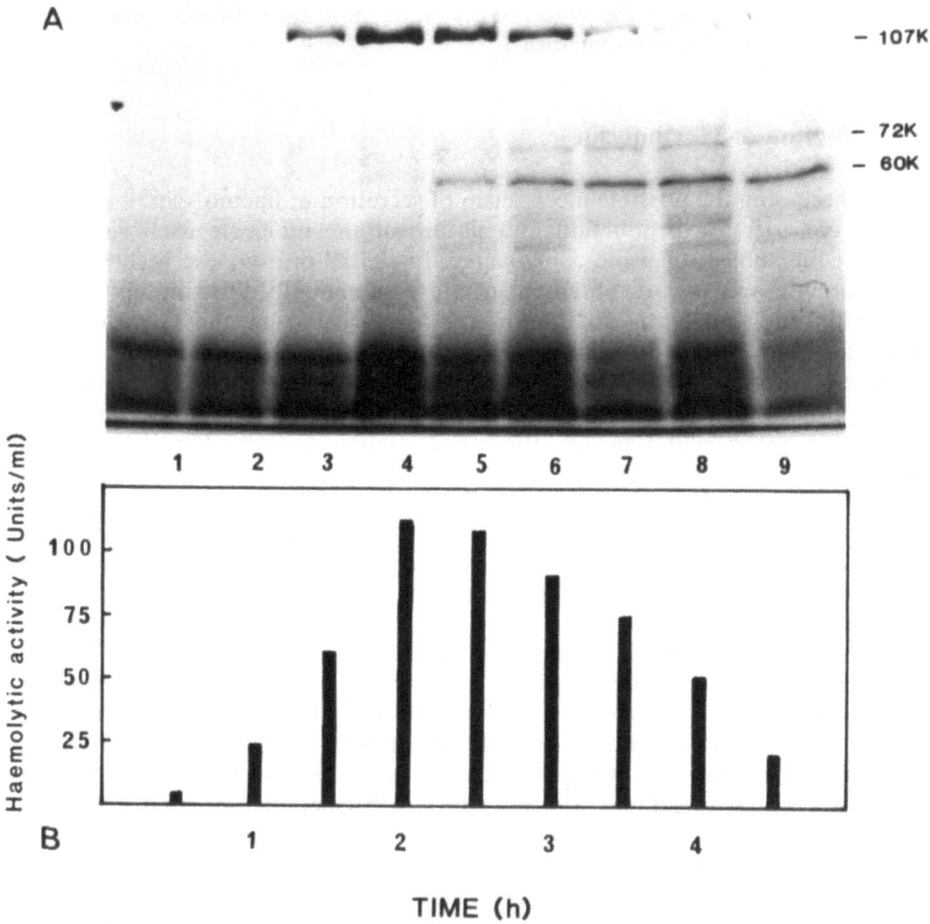

Fig. 2 A, B. Accumulation of the 107-kD polypeptide and external haemolytic activity during the growth cycle. Strain LE2001 was grown in Luria broth plus 10 m*M* calcium chloride. Ten-millilitre samples were taken at intervals during the growth cycle and external haemolytic activity determined **B**. In parallel supernatant proteins were analysed by SDS-PAGE (11% acrylamide) followed by staining with Coomassie blue (tracks 1–9, **A**). The 107-kD haemolysin and the major breakdown products are indicated

irreversibly to filters and dialysis tubing (unpublished data). In consequence, a range of molecular weights have been reported, 580000 (Rennie and Arbuthnott 1974), 120000 (Williams 1979) and 58000 (Noegel et al. 1979). These values were usually based upon gel filtration of samples purified using ion-exchange chromatography whilst monitoring for haemolytic activity. Crucially, however, these early reports failed to identify any specific proteins, e.g. by sodium dodecyl sulphate polyacrylamide gel electrophoresis (SDS-PAGE), in culture supernatants corresponding to the active haemolysin. Surprisingly, therefore, in 1984 we readily identified abundant amounts of a 107-kD polypeptide in the culture medium of many haemolytic strains of human and animal origin (MACKMAN and HOLLAND 1984a, b). In our preliminary purification studies (MACKMAN and HOLLAND 1984b), haemolytic activity was shown to cosediment with the 107-kD polypeptide in 10%–30% glycerol gradients and also to fractionate with this protein using ammonium sulphate precipitation.

These results were confirmed using an Hly determinant from an *E. coli* 04 serotype, strain LE2001, isolated from a human urinary tract infection. This Hly determinant was cloned into the 'runaway' vector pOU71 forming the recombinant plasmid pLG570. This vector has a copy number of one if maintained at 30° C (LARSEN et al. 1984). Furthermore, genetic analysis of this Hly determinant using Tn5 insertions into pLG570, which prevented the production of haemolysin, resulted in truncation of the 107-kD polypeptide. Other studies by NICAUD et al. (1985a), shown in Fig. 2, demonstrated that synthesis and stability of both the haemolytic activity and the 107-kD protein varied identically over a range of growth media and temperature conditions. Finally, the identity of the 107-kD polypeptide as haemolysin has recently been confirmed both by comparative protein and DNA sequence analysis (FELMLEE et al. 1985b) and by purification of the haemolysin to near homogeneity (GONZALEZ-CARRERO et al. 1985). This latter report also indicated that haemolysin was quite stable in the presence of 6 *M* urea. It is important to note that although none of these studies ruled out the possibility that the *active form* of the haemolysin molecule was a multimer or that additional low molecular weight components might be required for full activity (see below), recent studies have in fact shown that the active molecule is a monomer (S. Bhakdi, personal communication).

5 Structure of the Hly Determinant

Initial studies by GOEBEL and his coworkers (NOEGEL et al. 1979) established that the Hly determinant of the naturally occurring plasmid pHly152 comprised at least three genes which were essential for the synthesis and secretion of haemolysin. Attempts to clone the complete Hly determinant into a multicopy vector failed and cloning was only successful after fragmentation and then reassembly of the Hly determinant into a 'low-level expression site' in the multicopy vector pHSF1 (GOEBEL and HEDGPETH 1982). Thus, these workers isolated the haemolytic recombinant plasmid pANN202-312. Surprisingly this clone, although of higher copy number than the wild-type parental plasmid pHly152, produced at least ten times less haemolysin in the medium as determined by both haemolytic activity (MACKMAN and HOLLAND 1984b) and the amount of secreted 107-kD polypeptide (FELMLEE et al. 1985a; our unpublished data). In fact, minicells carrying pANN202-312 retained significant amounts of the 107-kD polypeptide (GOEBEL and HEDGPETH 1982), whilst maxicells carrying the wild-type plasmid pHly167 (MACKMAN and HOLLAND 1984a) and normal strains carrying either pHly152 or pHly167 (unpublished data) efficiently secreted the 107-kD protein to the medium. The basis of this apparent secretion defect in pANN202-312 has not yet been elucidated, although recently it has been reported (WELCH

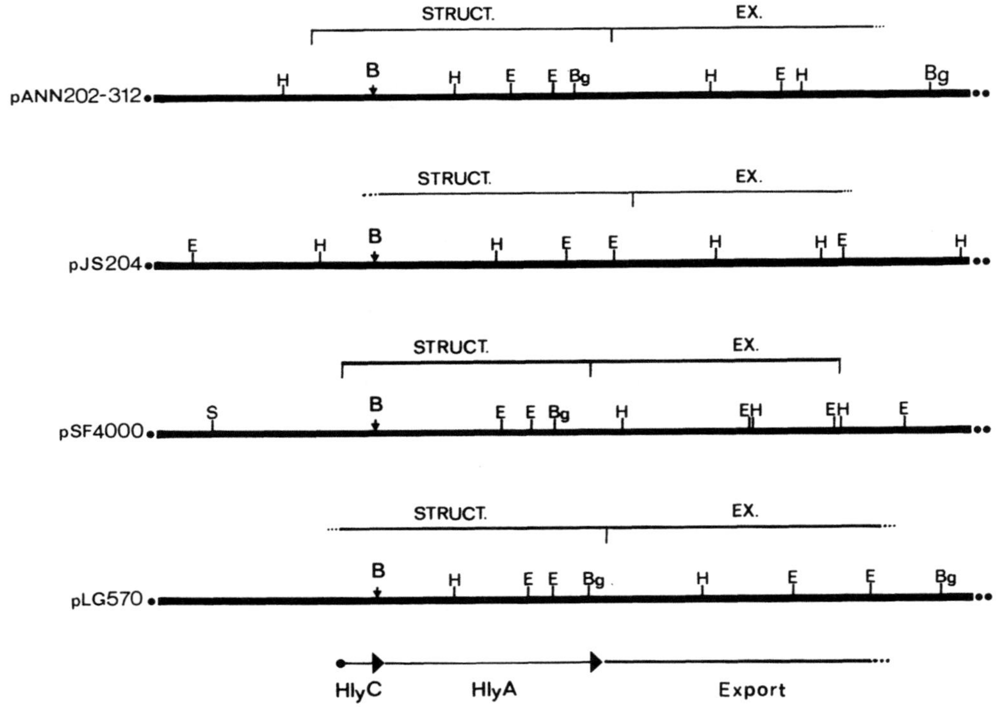

Fig. 3. Comparison of functional and restriction endonuclease maps of haemolysin determinants from various sources: pANN202-312 (Goebel and Hedgpeth 1982), pJS204 (no *Bgl*II mapping) (Stark and Shuster 1983), pSF4000 (Welch et al. 1983) and pLG570 (Mackman and Holland 1984b). The maps are arbitrarily aligned at the first *Bam*HI site. Recent sequence data have revealed that random base/pair mutations lead to changes in the restriction patterns of these determinants, although their overall sequences are essentially identical. *B*, *Bam*HI; *Bg*, *Bgl*II; *E*, *Eco*RI; *H*, *Hind*III; *struct*, structural genes; *ex*, export genes

and Falkow 1984) that insertion of an *Ava*I fragment from pSF4000, containing the promoter region for *hlyC*, in frame with the *hlyC* gene of pANN202-312 restores high levels of the extracellular 107-kD polypeptide. Thus, the clone pANN202-312 may contain some defect, for example, a down promoter mutation, compared with the original parental plasmid pHly152. In addition, the accumulation of the 107-kD polypeptide encoded by pANN202-312 inside minicells may suggest a further defect in the secretion mechanism.

Further studies of pANN202-312 (Wagner et al. 1983) nevertheless established by extensive transposon mapping and complementation studies that the complete Hly determinant was encoded by an approximately 8-kb DNA fragment. This consisted of at least four contiguous genes *hlyC*, *hlyA*, *hlyB* and *hlyD* (see Felmlee et al. 1985b for nomenclature) necessary for the production of active haemolysin in the medium. Several groups have now confirmed this general organization for a number of independently isolated Hly determinants by a variety of genetic and biochemical techniques (Stark and Shuster 1983;

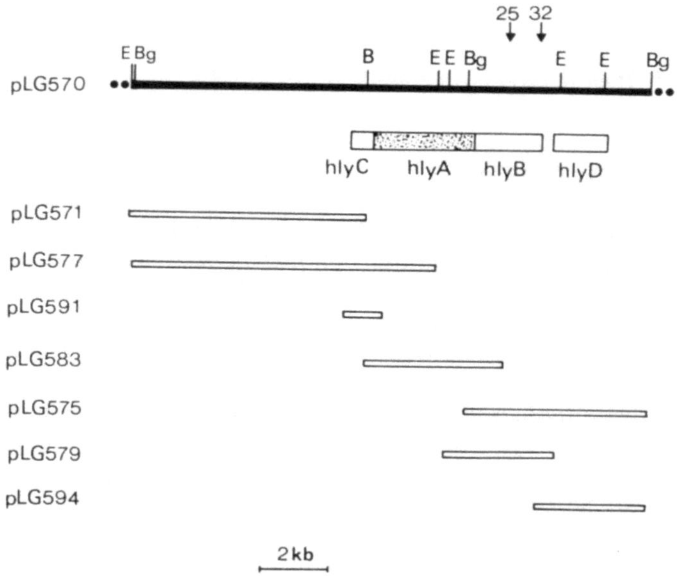

Fig. 4. Subclones of pLG570 carrying various Hly genes. All subclones, except pLG583 and pLG591, were constructed using the multicopy vector pACYC184 (for details see MACKMAN et al. 1985a). The location and isolate designation of two Tn5 insertions are shown by the *arrows above the horizontal bar*. In addition, pLG594 was formed by inserting the *Bgl*II fragment, carrying *hlyD* and part of Tn5 inserted at position 34 into pLG570, into the vector pACYC184

WELCH et al. 1983; MACKMAN and HOLLAND 1984b). The comparative restriction endonuclease maps are shown in Fig. 3 and indicate their close relationship.

Initial genetical studies (NOEGEL et al. 1979), which have now been confirmed by us (NICAUD et al. 1985b; MACKMAN et al. 1985a), showed that both *hlyC* and *hlyA* together were required for the production of haemolytic activity whereas *hlyB* and *hlyD* were essential for the secretion of the haemolysin. Thus, mutations in either *hlyB* or *hlyD* lead to accumulation of intracellular haemolytic activity. Figure 4 shows that appropriate subclones of the recombinant plasmid pLG570 now exist for the independent analysis of any of the four Hly genes. This has allowed us to reconstruct the complete system by introducing into *E. coli* strain CSH26 *rec*A three separate plasmids carrying *hlyC*, *hlyA*, and the combined export functions *hlyB* and *hlyD* respectively (MACKMAN et al. 1985a). Finally, recent DNA sequence analysis of an entire Hly determinant (from *E. coli* J96 serotype 04) has identified four open reading frames corresponding to polypeptide products of 20, 110, 80 and 54 kD respectively (FELMLEE et al. 1985b). Comparison of these data with those published for *hlyC* from pANN202-312 (JUAREZ et al. 1984) and our own sequence data of *hlyC* together with other parts of pLG570 (unpublished data) reveal a highly conserved gene structure. In fact, the two *hlyC* genes from pANN202-312 and pLG570 are remarkably homologous, with 97% of the DNA sequence being conserved (NICAUD et al. 1985b). Interestingly, there seems to be little homology upstream of *hlyC* in the region containing these *hlyC* promoters (unpublished data).

5.1 Identification of Promoters

Despite gross similarities in the organisation of Hly determinants studied in different laboratories, there remains some confusion in terms of possible promoters within the Hly determinant. Two groups (JUAREZ et al. 1984; MACKMAN et al. 1985a) have independently concluded, from studies of the polar effects on *hlyB* of transposon insertion into *hlyA* and by constructing Mu d1-directed *hly-lacZ* fusions, that *hlyC*, *hlyA* and *hlyB* form an operon which is transcribed from a promoter upstream of *hlyC*. In contrast, FELMLEE et al. (1985b) have provisionally suggested, from the DNA sequence and from the protein expressed by various subclones, that *hlyB* may be expressed from an independent promoter in vivo although this has not been confirmed by complementation studies. Nevertheless, in view of the strong polar effects of insertions in *hlyA* we suggest that *hlyB* is not actively transcribed from an independent promoter in vivo.

In the case of *hlyD*, DNA sequence data from pSF4000 (FELMLEE et al. 1985b) confirmed our studies with pLG570 (MACKMAN et al. 1985a) that *hlyD* was transcribed from a relatively strong independent promoter and that transcription was in the same direction as the *hlyC*, *hlyA*, *hlyB* operon (see Fig. 5a). This contrasts with the results reported by JUAREZ et al. (1984), in which the promoter for *hlyD* initiated transcription in the opposite direction to the transcription of *hlyC*, *hlyA* and *hlyB*.

5.2 Gene Organization in Hly2001

The gene organization of the Hly determinant 2001 present in pLG570 and studied extensively in this laboratory is summarized in Fig. 5. This structure was deduced from gene inactivation by transposon insertion coupled with complementation by specific subclones (shown in Fig. 4) to restore production of either intracellular or extracellular haemolytic activity. The various *hly* gene products were identified using minicells, maxicells and an in vitro coupled transcription-translation system (see below). Antibodies were also raised against the secreted 107-kD polypeptide, which allowed us to confirm the identification of truncated derivatives of the 107-kD protein as well as to localize intracellular haemolysin. As indicated in Fig. 5a, the Hly determinant is apparently transcribed from two promoters situated upstream of *hlyC* and *hlyD*. The general features of pLG570 have now largely been confirmed by FELMLEE et al. (1985b), although our transposon mutagenesis data has suggested a small intergenic region between *hlyB* and *hlyD* which does not appear to be present in pSF4000.

5.3 Characterization of Protein Products

hlyC: The *hlyC* gene product was first identified when the haemolytic plasmid pHly152 and various subclones were expressed in minicells (NOEGEL et al. 1979). A molecular weight of approximately 18000 was confirmed when the naturally occurring plasmid pHly167 was expressed in maxicells (MACKMAN and HOLLAND

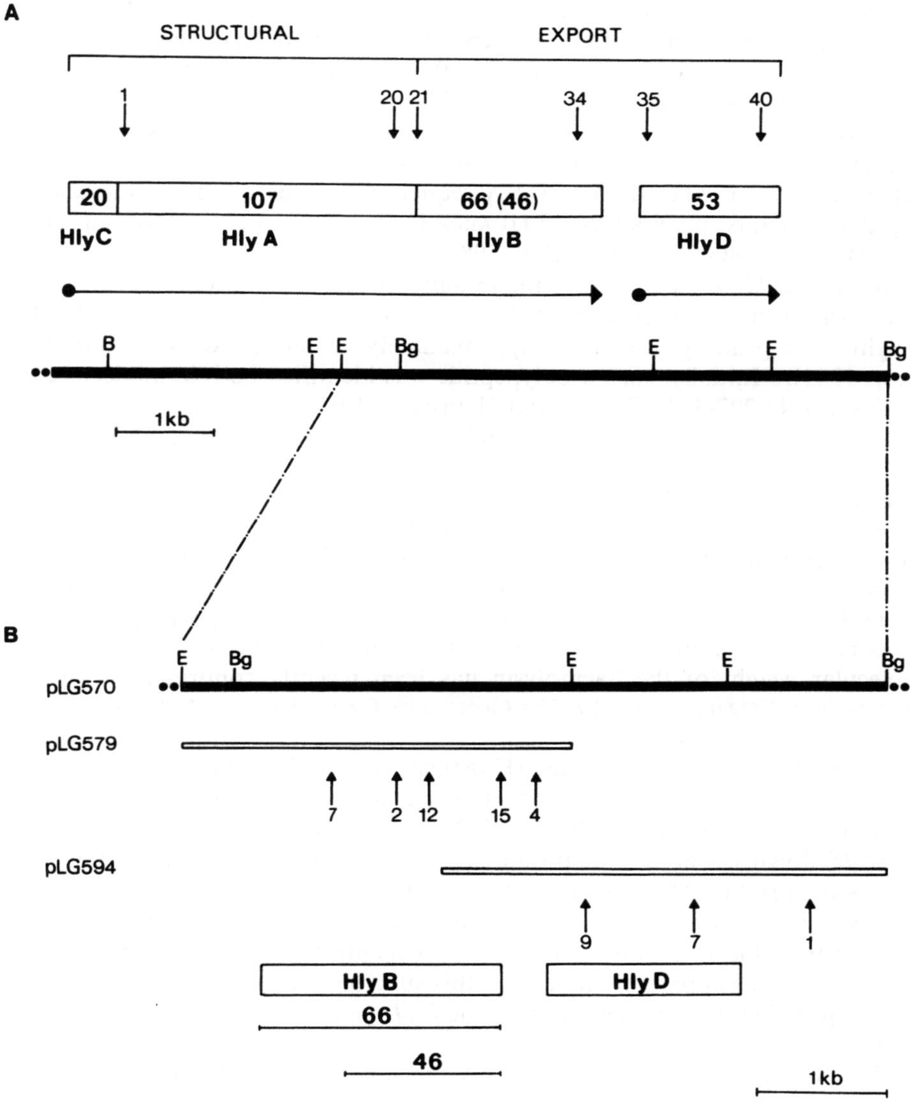

Fig. 5. A Genetic organization of the Hly determinant 2001. The location and isolate designations of several Tn5 insertions are shown by the *arrows above the horizontal bar*. Symbols for the restriction endonucleases are: *B*, *Bam*HI; *Bg*, *Bgl*II; *E*, *Eco*RI; *H*, *Hind*III. The two major promoters and the direction of transcription of the haemolysin genes are shown together with the extent of the various genes *hly*C, *hly*A, *hly*B and *hly*D drawn as *blocks below*. **B** Detailed analysis of the export region of pLG570. The various subclones were constructed as described in MACKMAN et al. (1985a). Subclones pLG579 and pLG594 carry *hly*B and *hly*D, respectively. The location of additional Tn5 insertions into these plasmids is shown by *arrows below the horizontal bars*. In pLG579, Tn5–7 to –15 were shown by complementation studies to map within *hly*B and in pLG594, Tn5–9 to –7 were shown to map within *hlyD*. The intergenic region between *hly*B and *hly*D, where Tn5 insertion –4 into pLG579 failed to disrupt *hly*B, is indicated. Other details, see text

1984a) and when the subclones pLG571 and pLG577 from pLG570 were expressed in vitro. In fact, pLG571 encoded a truncated *hlyC* gene product of 14 kD due to the construction of this subclone using the unique *Bam*HI site within the *hlyC* gene (see NICAUD et al. 1985b). Complementation studies using pLG571 (MACKMAN et al. 1985a) revealed that the 14-kD protein was inactive. The molecular weight of HlyC from these studies agrees very closely with a value of 20 kD predicted from DNA sequence analysis of three independent Hly determinants, pANN202-312 (JUAREZ et al. 1984), pLG570 (NICAUD et al. 1985b) and pSF4000 (FELMLEE et al. 1985b).

hlyA: The *hlyA* gene encodes the initially inactive haemolysin molecule and transposition mutagenesis indicated a gene size corresponding to a molecular weight for a primary product of approximately 107000 (MACKMAN and HOLLAND 1984b). Initially, such a polypeptide was identified both in minicells containing pANN202-312 (GOEBEL and HEDGPETH 1982) and in the medium from cells carrying the haemolytic plasmids pHly167 (MACKMAN and HOLLAND 1984a) or pLG570 (MACKMAN et al. 1985a). Moreover, in this laboratory insertion of Tn5 into *hlyA* generated a family of new polypeptides of largely the expected size. The *hlyA* gene from pLG570 has now been cloned separately to form pLG583 which encodes the 107-kD protein (see Fig. 6). In addition, subcloning an *Eco*RI fragment to form pLG577 (Fig. 4), which fails to produce haemolytic activity, programmes the synthesis of an 82-kD polypeptide which cross-reacts with antibody raised against the 107-kD haemolysin (Fig. 6). The molecular weight of the haemolysin has been recently confirmed by several groups (e.g. FELMLEE et al. 1985b; GONZALEZ-CARRERO et al. 1985, S. Bhakdi, personal communication).

HlyD: Goebel and co-workers (HARTLEIN et al. 1983) estimated the molecular weight of the *hlyD* gene product to be about 62 kD from expression in minicells of a subclone (pANN205-222) carrying the export functions (*hlyB* + *hlyD*) downstream of a *lac* promoter. However, the identity of this polypeptide as the product of *hlyD* was not confirmed. Studies in this laboratory with Hly2001 (Fig. 7) have shown that the subclone pLG575 carrying *hlyD* and *hlyB* programmed the synthesis of a 53-kD polypeptide both in minicells and in vitro, which was apparently unique to this subclone. Recently, by the construction of pLG594, which contains only the *hlyD* gene, we have shown the 53-kD protein to be the major product.

Finally, we have confirmed this as the *hlyD* product by Tn5 insertional mutagenesis which led to the loss of the 53-kD protein and the appearance of a 40-kD polypeptide. This is the molecular weight predicted by the site of Tn5 insertion (MACKMAN et al. 1985b). Recently, FELMLEE et al. (1985b) have identified an open reading frame corresponding to *hlyD* which predicts a polypeptide of 54-kD. In the same study, a polypeptide with this molecular weight was identified as HlyD by comparing the gene products expressed in minicells containing pANN202-312 with a subclone obtained by deleting the 5-kb *Bgl*II fragment containing *hylB* and *hlyD* (see Fig. 3).

hlyB: Characterization of the gene product of *hlyB* has proved to be more complex than the other *hly* genes because of its low level of expression. Firstly, a 46-kD polypeptide was identified when a subclone (pANN205-222) of the

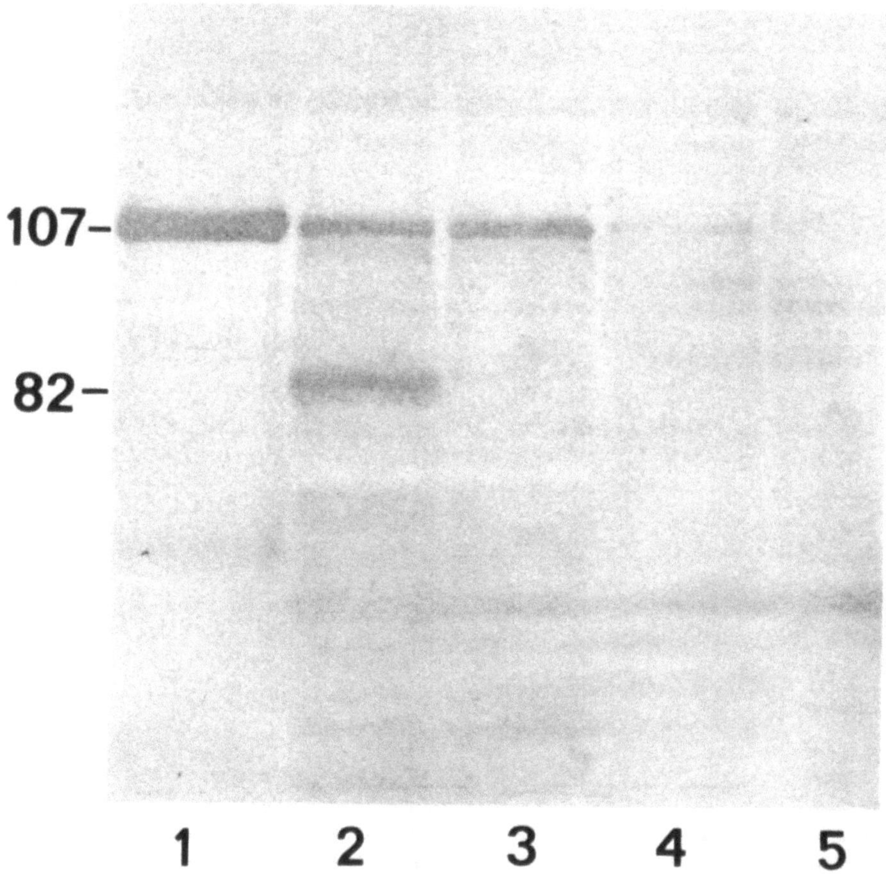

Fig. 6. Immunodetection of the 107-kD *hly*A gene product. Cultures of *E. coli* K12 MC4100 containing various plasmids were grown to mid-exponential phase before harvesting and solubilizing the cell proteins in SDS sample buffer. These were separated by SDS-PAGE (11% acrylamide), transferred to nitrocellulose filters and detected using antibodies raised against purified 107-kD polypeptide. Track 1, purified 107-kD polypeptide standard; track 2, MC4100 pLG577/pLG583; track 3, MC4100 pLG583; track 4, MC4100 pLG570; track 5, MC4100

export system of pANN202-312 was expressed in minicells (HARTLEIN et al. 1983). A protein of similar size is also observed when the *hlyB* gene from pLG570 is expressed in an in vitro coupled transcription translation system (Fig. 7). Secondly, a large polypeptide with a molecular weight of 77-kD has been reported by FELMLEE et al. (1985b) to be expressed in minicells by an *EcoRI* fragment (pWAM326 and pWAM327) containing *hlyB*. Since inspection of the DNA sequence indicated the possibility of an internal translation start within the open reading frame corresponding to *hlyB*, FELMLEE et al. (1985b) proposed that the 46-kD protein was translated from within the larger region of DNA encoding the 77-kD polypeptide.

In extensive studies with pLG570 (see Fig. 7) we have shown that the major product encoded by a subclone (pLG579) carrying *hlyB* in minicells was in

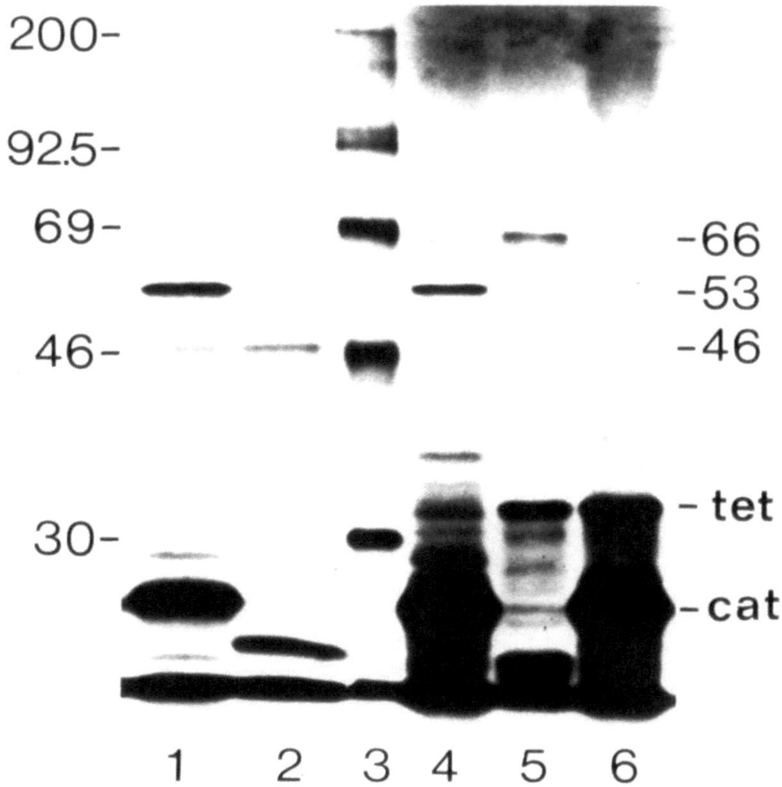

Fig. 7. Gene products encoded by *hly*B and *hly*D. Proteins were labelled with [^{35}S]-methionine and visualised by autoradiography after SDS-PAGE (11% acrylamide). Plasmids pLG575 (track 1) and pLG579 (track 2) were expressed in the in vitro coupled transcription-translation system (PRATT et al. 1981). In addition, pLG575 (track 4), pLG579 (track 5) and vector pACYC184 (track 6) were expressed in minicells. Track 3 shows the ^{14}C standards with molecular weights in kilodaltons. The 66-kD, 46-kD *hlyB* gene products and the 53-kD *hlyD* protein are indicated. See text for details

fact a 66-kD polypeptide. In contrast, when the same *hlyB* subclone was expressed in vitro the 46-kD protein was the major product (Fig. 7), with only small amounts of the 66-kD protein visible. In order to establish the precise relationship between these polypeptides and the *hlyB* gene transposon mutagenesis was carried out on pLG579 using Tn5. Firstly, the transposition inactivation map indicated the complete size of *hlyB* to be sufficient to encode a polypeptide of about 70-kD. Secondly, analysis of specific Tn5 insertions (e.g. pLG579:Tn5–12 – see Fig. 5b) into *hlyB* expressed in both minicells or in vitro showed the elimination of the 66-kD and 46-kD polypeptides and the appearance of the predicted 56-kD and 36-kD truncates respectively. Therefore, by comparing these truncated polypeptides with the respective map positions of the Tn5 insertions, we also conclude that the 66-kD and 46-kD polypeptides correspond to overlapping translational frames shown in Fig. 5b (MACKMAN et al. 1985b). We do not, however, know whether both of these polypeptides are essential for secretion of HlyA. In pulse-chase experiments using minicells containing pLG579 we were unable to demonstrate any precursor-product relationship between the 66-kD and 46-kD proteins.

These transposon inactivation studies also suggest the presence of an approximately 200-bp intergenic region between *hlyB* and *hlyD* (Fig. 5b) but this has yet to be confirmed. This is not present in the DNA sequence data of FELMLEE et al. (1985b), although sequence data from this laboratory in this region just upstream of *hlyD* agree with the published data. A smaller *hlyB* region as found with pLG570 would of course be more consistent with the size of polypeptides so far proposed to be encoded by *hlyB*. Moreover, the molecular weight of 66000 estimated by SDS-PAGE may be anomalously low. In summary, the polypeptide products of the Hly determinant 2001 and the major Tn5 insertions used in the above studies are shown in Fig. 5.

Finally, we would like to emphasize that in our hands identification of *hlyB* products, which are expressed at very low levels, was only possible using the smallest available subclones. Furthermore, we consider it essential that the identity of both HlyB and HlyD be confirmed by demonstrating the formation of truncated polypeptides by Tn5 insertions which inactivate the genes.

5.4 Role of HlyC in the Production of Active Haemolysin

As indicated above in addition to *hlyA*, the *hlyC* gene product is also required in some way for the production of an active intracellular haemolysin. NOEGEL et al. (1979) proposed that HlyC was required for 'activation' of the haemolysin and also for the passage of haemolysin across the inner membrane to the periplasm prior to its final translocation to the medium. However, studies with the Hly determinant 2001 have now shown conclusively that the presence of recombinant plasmids pLG583 and pLG575 (see Fig. 4), as a source of HlyA and the export functions respectively, is sufficient for efficient secretion of an inactive 107-kD polypeptide to the medium (NICAUD et al. 1985b). HlyC therefore has no role in secretion of the haemolysin. These experiments also demonstrated that HlyC is not required for synthesis of HlyA. This leaves two possible roles for HlyC, either HlyC directly modifies HlyA post-translationally to produce the active form or HlyC combines with the 107-kD molecule in the medium to form an active complex. Cellular fractionation studies have consistently failed to detect HlyC in the culture medium. In fact, the majority of the molecule remains exclusively with the soluble, membrane-free fraction when cells are disrupted by sonication (unpublished data) and we propose that it is primarily localized in the cytoplasmic fraction. Moreover, our recent studies (NICAUD et al. 1985b and unpublished data) have shown that unmodified HlyA isolated from the culture medium migrates slower in non-denaturing PAGE, is more stable and shows differential resistance to staphylococcal V8 protease compared with the active form. Both active and inactive forms are, however, identical in molecular weight as determined by SDS-PAGE. We conclude that HlyC confers some form of post-translational modification upon the 107-kD polypeptide which 'activates' the molecule. Preliminary analyses indicate that this modification does not involve extensive glycosylation or phosphorylation of the molecule (unpublished data).

5.5 Localization of the Export Proteins

The localization of the 'export' proteins HlyB and HlyD is vital to an understanding of the precise mechanism of secretion of the haemolysin. Unfortunately the export functions, particularly *hlyB*, are expressed at a very low level and normally cannot be detected in whole cells (NICAUD et al. 1985c). Consequently, localization studies have been restricted to the use of minicells or maxicells and only a limited range of membrane fractionation procedures can be applied. Previously, HARTLEIN et al. (1983) have reported that the 46- and 62-kD polypeptides, suggested by these workers to be the products of *hlyB* and *hlyD* respectively, were primarily localized to the outer membrane fraction of minicells when envelopes were separated on Osborn sucrose gradients (OSBORN et al. 1972). Unfortunately, few experimental details of this fractionation procedure, which is extremely difficult to carry out with minicells, were given and it is therefore difficult to evaluate these results.

Preliminary localization studies have been carried out in this laboratory using the export subclone pLG575 containing *hlyB* and *hlyD* from pLG570 expressed in minicells and maxicells (MACKMAN et al. 1985b). In view of the difficulties alluded to above, these fractionation studies have so far been limited to sarkosyl treatment of envelopes isolated from sonicated cells to separate the soluble (inner membrane) fraction from the insoluble (outer membrane) fraction (FILIP et al. 1973). When known inner membrane (e.g. PBP5) and outer membrane (TonA) proteins are used as controls (MACKMAN et al. 1985b) minicells and maxicells gave essentially the same results as normal *E. coli* K12 strains. The 53-kD, *hlyD* protein clearly sedimented with the envelopes from both minicells and maxicells and fractionated primarily with the inner membranes although small amounts were also detected in the sarkosyl insoluble fraction. Preliminary data also suggest that the majority of the low levels of *hlyB* gene product(s) is located in the inner membrane. Interestingly, inspection of the DNA sequence of *hlyD* (FELMLEE et al. 1985b) indicates the presence of one major hydrophobic domain in the corresponding protein which might form a transmembrane region. Several such domains were apparently present in HlyB. Surprisingly, the sequence data do not, however, indicate the presence of any N-terminal signal sequences for either the 53-kD or the 66-kD proteins. This is also more consistent with these proteins being in the inner membrane rather than the outer membrane. Nevertheless, further fractionation data to confirm the location, preferably in whole cells, are still required since the basis of membrane protein solubility in sarkosyl is not known and the behaviour of HlyB and HlyD in this detergent cannot be assumed. However, as discussed below, there may be other data concerning the function of these proteins which are more consistent with an inner membrane location for these haemolysin export proteins (see Sect. 7).

6 Structure of HlyA Relevant to the Secretion Mechanism

Hartlein et al. (1983) previously suggested that the HlyA protein does not contain an N-terminal signal sequence. This has now been confirmed by Felmlee et al. (1985a), who showed that the N-terminal amino acid sequence of the extracellular 107-kD protein is identical to that predicted by the DNA sequence of *hlyA*. Moreover, the DNA sequence data indicate that the HlyA molecule is remarkably polar, with a large number of basic residues in the N-terminal region and many acidic residues in the C-terminal region. Since the molecule, like several colicins (Holland 1975), is highly polar and does not contain any cysteine residues, we speculate that it may be an elongated molecule rather than a globular protein. The DNA sequence, except for the prediction of three rather hydrophobic domains between amino acid residues 230 and 405 of the protein, gives few clues as to its mode of secretion. However, it is unlikely that this will depend upon the *secA*, *prlA* pathway which probably involves recognition of an N-terminal signal sequence (Oliver and Beckwith 1981). More likely, the key to understanding the export of haemolysin will lie in the presumed recognition of HlyA by the specific export machinery, the HlyB and HlyD proteins. As an initial approach to this we have now isolated mutations in *hlyA* which result in failure to secrete the haemolysin.

7 A Periplasmic Intermediate in Secretion of Haemolysin?

It is generally considered that secretion of proteins in other Gram-negative organisms is a two-step process involving translocation to the periplasm followed by transfer across the outer membrane (see also Pugsley and Schwartz 1985). This process may be analogous to the assembly of fimbriae on the surface of *E. coli* (Mooi et al. 1983). The demonstration of a periplasmic intermediate in the secretion of haemolysin could therefore be an important step in establishing the nature of the export process.

Surprisingly, from initial studies by Wagner et al. (1983) on the haemolytic activity encoded by both pHly152 and pANN202-312, it was reported that 50% of the total haemolytic activity was located in the periplasm during *normal* secretion, with 30% in the medium and only 10% in the cytoplasm. Furthermore, in minicells lacking either of the export functions, 70% of the haemolytic activity was apparently present in the periplasm compared with 20% in the cytoplasm (Hartlein et al. 1983). However, it was also claimed that 'an inactive' 107-kD polypeptide (synthesised in the absence of *hlyC*), as shown by cell fractionation of minicells, remained exclusively in the cytoplasm. Thus, it was concluded that the secretion of haemolysin was a two-step process involving a periplasmic intermediate and that the HlyC protein was somehow essential for transport across the inner membrane.

These results obviously conflict with our findings discussed above that the 107-kD polypeptide encoded by pLG583 is efficiently secreted to the medium in the complete absence of *hlyC*. Moreover, only low levels of 107-kD polypep-

Table 1. Accumulation of haemolytic activity

Plasmid	Supernatant	%	Periplasm	%	Cytoplasm	%
pLG570	160	97	0.3	0	3.8	3
pLG583/ pLG577	0.11	0	1.7	9	17.6	90

Strain MC4100 carrying the various plasmids was grown in nutrient broth and harvested at optical density $(OD)_{450}=1.2$. Cells were fractionated as described by NOSSAL and HEPPEL (1966) and units of haemolytic activity determined as defined in MACKMAN and HOLLAND (1984a, b)

tide are detected inside the cells under these conditions using antiserum (Fig. 6, track 4). In view of these results, the fractionation data with minicells containing pANN202-312 export mutants should be treated with some caution. As indicated above, secretion of haemolysin by pANN202-312 is in any case very inefficient and therefore this may not be an ideal system in which to study the secretion process.

Accumulation of haemolytic activity within the periplasm of bacteria carrying pLG570 with mutations in the export functions has also been reported previously (MACKMAN and HOLLAND 1984b). However, we have re-examined this question in some detail involving studies on several mutations in either *hlyB* or *hlyD* as well as a system where both functions are completely lacking. In order to minimize possible difficulties in cellular fractionation, we chose to use exponentially growing cultures of the K12 strains MC4100 and CSH26 rather than mini or maxicells. The results are presented in Table 1. In contrast to the fractionation data obtained by WAGNER et al. (1983), strains containing pLG570 secreted 97% of total haemolytic activity into the medium, with the remainder in the cytoplasm and no significant levels of activity in the periplasm. In export-defective mutants all the activity remained with the cells but strikingly the maximum levels obtained were only 10% of that normally exported to the medium with the intact system. When the distribution of intracellular activity was examined, Table 1 shows that in the complete absence of both export functions 90% of the activity was in the cytoplasm. In the absence of either *hlyB* or *hlyD* the majority of haemolytic activity was present in the cytoplasm but once more the levels only represented about 5% of total activity produced by pLG570 (unpublished). The significance of the intracellular distribution of such low levels of activity is difficult to assess, especially since even with the most careful regime the crossover level of the cytoplasmic protein malate dehydrogenase and the periplasmic protein β-lactamase is about 10%. For these reasons we suggest, in contrast to our previous conclusion, that if there is a normal periplasmic intermediate in haemolysin secretion it remains to be demonstrated.

Since the export mutants displayed very low levels of haemolytic activity it appears that intracellular haemolysin is either extremely unstable or alternatively that synthesis of HlyA is somehow switched off. This is currently being investigated.

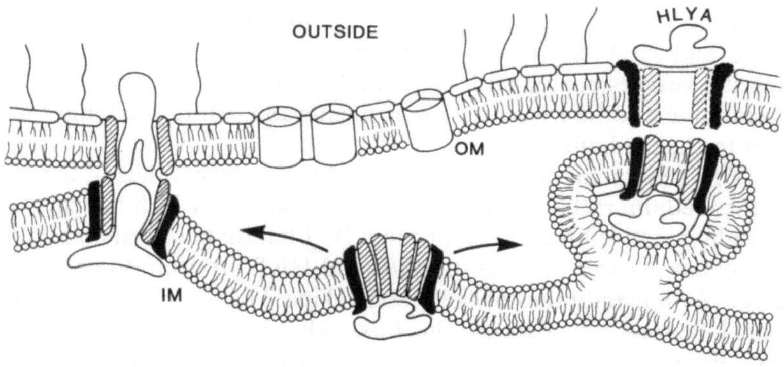

Fig. 8. Mechanism of secretion of haemolysin from *E. coli.* The model illustrates the possible extrusion of HlyA directly to the medium (*on the left*) through specific regions of adhesion between inner and outer membranes. This is proposed to be facilitated by the export proteins HlyB and HlyD, which are largely located within the inner membrane. The model emphasizes a specific interaction between HlyA and one or both of the export proteins leading to a proposed energy-dependent translocation of HlyA to the exterior. *On the right* is indicated a variation on the model whereby HlyB and HlyD, initially localized to the inner membrane, interact with HlyA, with the concomitant budding of a vesicle which ultimately fuses with the outer membrane releasing HlyA directly to the medium

8 Possible Models for Haemolysin Export

With the recent accumulation of considerably more data on the nature of the haemolysin secretion system, it now seems necessary to formulate a new model to accommodate the additional findings. Firstly, however, it is important to review briefly the major model outlined previously by Goebel and his co-workers (Hartlein et al. 1984). This model was based upon the identification of a 58-kD polypeptide as haemolysin, cellular fractionation of haemolytic activity and localization of the export proteins to the outer membrane. These results were mainly derived from expression of the Hly determinant pANN202-312 in minicells (Hartlein et al. 1983). Thus, the following model was proposed. The primary product of *hlyA*, a 107-kD polypeptide, is proteolytically processed by the *hlyC* product; as a consequence the 58-kD protein, now haemolytically active, is translocated across the inner membrane to the periplasm. The presence of the HlyB and HlyD proteins in the outer membrane then promotes translocation of the 58-kD haemolysin to the medium.

On the basis of more recent studies, particularly involving pLG570 in this laboratory, we now suggest a substantially different mechanism, illustrated in Fig. 8. First, the 107-kD, *hlyA* gene product is secreted to the medium irrespective of the presence of *hlyC* which is only required for the activation of HlyA. Secondly, interaction of HlyA with HlyB and/or HlyD, which from preliminary results appear to be primarily bound to the inner membrane, promotes the passage of HlyA across the inner membrane. Thirdly, in the absence of strong evidence for a periplasmic intermediate in HlyA secretion, we propose the more economical hypothesis that HlyA is then directly extruded to the medium. This might be achieved either by the formation of specific export sites formed by HlyB and HlyD at 'Bayer-like junctions', or through entrapment of HlyA in specific vesicles coated with HlyB and/or HlyD which then bud from the inner membrane and fuse with the outer membrane before shedding their lumen content to the medium.

This general model first of all explains the appearance of the 107-kD protein in the medium rather than the 58-kD polypeptide, which appears to be a haemolytically inactive breakdown product (S. Bhakdi, personal communication). The model emphasises the role of HlyB and HlyD in promoting transport of HlyA across *both* membranes of the cell envelope. However, the participation of additional factors provided by the host cell for the export process is not ruled out. Finally, the model makes no distinction at this stage between the roles of the HlyB protein(s) and HlyD in secretion and both may well form a specific complex which is recognised by HlyA.

9 Perspectives

The basic structure of the *E. coli* haemolysin determinant is now clear. In particular, great progress has been made as a result of the demonstration, now widely confirmed, that the haemolytic factor secreted to the medium is the 107-kD polypeptide. Furthermore, the identity of the export proteins HlyB and HlyD is now broadly established. On the other hand, the precise mechanism by which *E. coli* secretes to the medium at least 0.1% of its total protein (i.e. at least 10^4 molecules per cell per generation) in the form of this high molecular weight, highly polar polypeptide is still completely unclear. Even the origin of the Hly determinant as a pathogenic factor is uncertain as its DNA composition suggests an original host unrelated to *E. coli* (Felmlee et al. 1985b).

The central question now to be addressed is the mechanism which allows the 107-kD haemolysin to negotiate two hydrophobic membrane layers on its way to the medium without disturbing the growth of the organism in any detectable way. The key to the understanding of this mechanism appears to reside in the nature of the interaction (now vital to establish) between the 107-kD protein and the membrane-bound 66-kD (and 46-kD) and 53-kD export proteins. This will require the unequivocal localization of the export proteins within the cell envelope. In turn, it will be necessary to construct plasmids which will allow the controlled overproduction and purification of these polypeptides and, for example, to attempt the reconstitution of the secretion process in vitro.

Not least of several fascinating aspects of haemolysin production is the absence of N-terminal signal sequences in either HlyA itself or in the 53-kD and 66-kD export proteins. Establishing the nature of the alternative mechanism, involving possible topogenic sequences which circumvent the need for an N-terminal signal, is an intriguing prospect which may nevertheless allow a better understanding of the more conventional export process itself. Indeed, although all studies so far have concentrated upon the role of Hly determinants in the secretion process, some involvement of *E. coli* biochemical and physiological systems in the biogenesis of cell envelope components is highly likely. At least one *E. coli* chromosomal mutation has been reported which blocks secretion of haemolysin (JUAREZ and GOEBEL 1984) and preliminary studies in this laboratory have also indicated the importance of other *E. coli* genes. It is highly probable that translocation of the 107-kD polypeptide through one or both membranes will be energy dependent and this is but one aspect where this novel secretion mechanism is likely to impinge upon the physiology of *E. coli*.

Other features of the haemolysin system still requiring clarification include the nature of growth phase regulation, with the possibility of feedback control between synthesis and the export process; the relationship of such regulatory mechanisms and overall control of surface biosynthesis in the *E. coli* host which itself is still poorly understood. All these questions are now amenable to resolution by recombinant DNA techniques. Similarly, the relationship between structure and function within the HlyA protein is now an exciting prospect available for study. Finally, the identification of the 107-kD protein as haemolysin and its availability in highly purified form will facilitate the elucidation of its mode of action, in particular its significance as a cytotoxic factor, in addition to its haemolytic activity, in the pathogenic process. An integral part of this process is likely to be the nature of the post-translational modification promoted by the HlyC protein. Establishing the nature of this modification and its precise contribution to the stability and/or interaction with the surface of mammalian cells is urgently required.

Acknowledgements. This work was supported by grant no GR/C81148 and a CEC grant from the Biomolecular Engineering Programme. We would like to thank P. Williams for reading the manuscript and S. Bhakdi for discussion of unpublished data. Finally, we would like to acknowledge R. Welch for some very helpful discussions.

References

Andro T, Chambost J-P, Katoujansky A, Cattaneo J, Bertheau Y, Barras F, Van Gijsegem F, Coleno A (1984) Mutants of *Erwinia chrysanthemi* defective in the secretion of pectinase and cellulase. J Bacteriol 160:1199–1203

Bayer ME (1979) The fusion sites between outer membrane and cytoplasmic membrane of bacteria. In: Inouye M (ed) Bacterial outer membranes. Willey, New York, pp 167–202

Caro LA, Palade GE (1964) Protein synthesis, storage and discharge in the pancreatic exocrine cell: an autoradiographic study. J Cell Biol 20:473–495

Cavalieri SJ, Snyder IS (1982a) Cytotoxic activity of partially purified *Escherichia coli* alpha haemolysin. J Med Microbiol 15:11–21

Cavalieri SJ, Snyder IS (1982b) Effect of *Escherichia coli* alpha-hemolysin on human peripheral leukocyte viability in vitro. Infect Immun 36:455–461

Cavalieri SJ, Bohach GA, Snyder IS (1984) *Escherichia coli* α-haemolysin: characteristics and probable role in pathogenicity. Microbiol Rev 48:326–343

Cavard D, Bernadac A, Lazdunski C (1981) Exclusive localisation of colicin A in the cell cytoplasm of producing bacteria. Eur J Biochem 118:125–131

Charbit A, Clement J-M, Hofnung M (1984) Further analysis of the phage lambda receptor site. Possible implications for the organization of the LamB protein in *Escherichia coli* K-12. J Mol Biol 174:395–401

Copeland BR, Richter RJ, Furlong CE (1982) Renaturation and identification of periplasmic proteins in two-dimensional gels of *Escherichia coli*. J Biol Chem 257:15065–15071

Dodd DD, Bassford PJ, Eisenstein BI (1984) Dependence of secretion and assembly of Type I fimbrial subunits of *E. coli* on normal protein export. J Bacteriol 159:1077–1079

Felmlee T, Pellett S, Lee E-Y, Welch R (1985a) *Escherichia coli* haemolysin is released extracellularly without cleavage of a signal peptide. J Bacteriol 163:88–93

Felmlee T, Pellett S, Welch R (1985b) Nucleotide sequence of an *Escherichia coli* chromosomal hemolysin. J Bacteriol 163:94–105

Filip C, Fletcher G, Wulff JL, Earhart C (1973) Solubilization of the cytoplasmic membrane of *Escherichia coli* by the ionic detergent sodium lauryl sarcosinate. J Bacteriol 115:717–722

Gadeberg OV, Orskov I (1984) In vitro cytotoxic effect of α-hemolytic *Escherichia coli* on human blood granulocytes. Infect Immun 45:255–260

Goebel W, Hedgpeth J (1982) Cloning and functional characterization of the plasmid-encoded hemolysin determinant of *Escherichia coli*. J Bacteriol 151:1290–1298

Gonzalez-Carrero MI, Zabala JC, de la Cruz F, Oritz JM (1985) Purification of α-haemolysin from an overproducing *E. coli* strain. MGG 199:106–110

Hall MN, Silhavy TJ (1981) Genetic analysis of the major outer membrane proteins of *Escherichia coli*. Annu Rev genet 15:91–142

Hartlein M, Schiessl S, Wagner W, Rdest U, Kreft J, Goebel W (1983) Transport of haemolysin by *Escherichia coli*. J Cell Biochem 22:87–97

Hartlein M, Hughes C, Muller D, Kreft J, Goebel W (1984) Haemolysin genes from Gram-negative and Gram-positive bacteria. In: Alouf JE, Fehrenbach FJ, Freer JH, Jeljaszewicz J (eds) Bacterial protein toxins. Academic, London, pp 39–46 (FEMS symposium no 24)

Hobot JA, Carlemalin E, Villiger W, Kellenberger E (1984) Periplasmic gel: new concept resulting from the re-investigation of bacterial cell envelope ultrastructure by new methods. J Bacteriol 160:143–152

Holland IB (1975) Physiology of colicin action. Adv Microbiol Physiol 12:55–139

Howard SP, Buckley JT (1983) Intracellular accumulation of extracellular proteins by pleiotropic export mutants of *Aeromonas hydrophila*. J Bacteriol 154:413–418

Jorgenson S, Short E, Kurtz H, Mussen H, Wu G (1976) Origins of a-haemolysin produced by *E. coli*. J Med Microbiol 9:173–189

Jorgenson S, Hammer RF, Mussen H, Wu GK (1980) Effects of a single hit from the alpha haemolysin produced by *Escherichia coli* on the morphology of sheep erythrocytes. Infect Immun 27:988–994

Juarez A, Hughes C, Vogel M, Goebel W (1984) Expression and regulation of the plasmid-encoded hemolysin determinant of *Escherichia coli*. MGG 197:196–203

Larsen H, Gerdes K, Light J, Molin S (1984) Low copy number plasmid vectors amplifiable by derepression of an inserted foreign promoter. Gene 28:45–54

Lugtenberg B, Van Alphen L (1983) Molecular architecture and functioning of the outer membrane of *Escherichia coli* and other Gram-negative bacteria. Biochem Biophys Acta 737:51–115

Mackman N, Holland IB (1984a) Secretion of a 107K dalton polypeptide into the medium from a haemolytic *E. coli* K12 strain. MGG 193:312–315

Mackman N, Holland IB (1984b) Functional characterization of a cloned haemolysin determinant from *E. coli* of human origin, encoding information for the secretion of a 107K polypeptide. MGG 196:123–134

Mackman N, Nicaud J-M, Gray L, Holland IB (1985a) Genetical and functional organisation of the *Escherichia coli* haemolysin determinant 2001. MGG 201:282–288

Mackman N, Nicaud J-M, Gray L, Holland IB (1985b) Identification of polypeptides required for the export of haemolysin 2001 from *E. coli*. MGG 201:529–536

Mooi FR, Wijfes A, de Graaf FR (1983) Identification and characterization of precursors in the biosynthesis of K88ab fimbria of *Escherichia coli*. J Bacteriol 154:41–49

Müller D, Hughes C, Goebel W (1983) Relationship between plasmid and chromosomal haemolysin determinants of *Escherichia coli*. J Bacteriol 153:846–851

Nicaud J-M, Mackman N, Gray L, Holland IB (1985a) Regulation of haemolysin synthesis in *E. coli* determined by Hly genes of human origin. MGG 199:111–116

Nicaud J-M, Mackman N, Gray L, Holland IB (1985b) Characterization of *HlyC* and mechanism of activation and secretion of haemolysin from *E. coli* 2001. FEBS letters 187:339–344

Nicaud J-M, Mackman N, Holland IB (1985c) Amplification of synthesis and secretion of haemolysin using a run-away plasmid in *E. coli*. J Biotech 3:175–183

Nikaido H, Nakae T (1979) The outer membrane of gram-negative bacteria. Adv Microb Physiol 20:163–250

Noegel A, Rdest U, Springer W, Goebel W (1979) Plasmid cistrons controlling synthesis and excretion of the exotoxin α-haemolysin of *Escherichia coli*. MGG 175:343–350

Nossal NG, Heppel LA (1966) The release of enzymes by osmotic shock from *Escherichia coli* in exponential phase. J Biol Chem 241:3055–3062

Oliver DB, Beckwith J (1981) *E. coli* mutants pleiotropically defective in the export of secreted proteins. Cell 25:765–772

Osborn MJ, Wu HCP (1980) Proteins of the outer membrane of gram-negative bacteria. Annu Rev Microbiol 34:369–422

Osborn MJ, Gander JE, Parisi E (1972) Mechanism of assembly of the outer membrane of *Salmonella typhimurium*: site of synthesis of lipopolysaccharide. J Biol Chem 247:3973–3986

Pratt J, Boulnois G, Darby V, Orr E, Wahle E, Holland IB (1981) Identification of gene products programmed by restriction endonuclease DNA fragments using an *E. coli in vitro* system. Nucl Acids Res 9:4459–4474

Pugsley AP, Schwartz M (1984) Colicin E2 release: lysis, leakage or secretion? Possible role of a phospholipase. EMBO J 3:2393–2397

Pugsley AP, Schwartz M (1985) Export and secretion of proteins by bacteria. FEMS Microbiol Rev 32:3–38

Rennie RP, Arbuthnott JP (1974) Partial characterization of *Escherichia coli* haemolysin. J Med Microbiol 7:179–188

Smith HW (1963) The haemolysins of *Escherichia coli*. J Pathol Bacteriol 85:197–211

Snyder IS, Koch NA (1966) Production and characteristics of haemolysin of *Escherichia coli*. J Bacteriol 91:763–767

Springer W, Goebel W (1980) Synthesis and secretion of haemolysin by *Escherichia coli*. J Bacteriol 144:53–59

Stark JM, Shuster CW (1983) The structure of cloned hemolysins DNA from plasmid pHly185. Plasmid 10:45–54

Talmadge K, Gilbert W (1982) Cellular location affects protein stability in *Escherichia coli*. Proc Natl Acad Sci USA 79:1830–1833

Talmadge K, Stahl S, Gilbert W (1980) Eukaryotic signal sequence transports insulin antigen in *Escherichia coli*. Proc Natl Acad Sci USA 77:3369–3373

Wagner W, Vogel M, Goebel W (1983) Transport of hemolysin across the outer membrane of *Escherichia coli* requires two functions. J Bacteriol 154:200–210

Welch RA, Falklow S (1984) Characterization of *Escherichia coli* hemolysins conferring quantitative differences in virulence. Infect Immun 43:156–160

Welch RA, Patchen-Dellinger E, Minshew B, Falklow S (1981) Haemolysin contributes to virulence of extraintestinal *E. coli* infections. Nature 294:665–667

Welch RA, Hall R, Falklow S (1983) Molecular cloning and physical characterization of a chromosomal haemolysin from *Escherichia coli*. Infect Immun 42:178–186

Williams P (1979) Determination of the molecular weight of *Escherichia coli* α-haemolysin. FEMS Microbiol Lett 5:21–24

Zabala JC, Garcia-Lobo JM, Diaz-Aroca E, de la Cruz F, Ortiz JM (1984) *Escherichia coli* alpha-haemolysin synthesis and export genes are flanked by a direct repetition of IS91-like elements. MGG 197:90–97

Production and Release of Cloacin DF13 and Related Colicins

F.K. DE GRAAF and B. OUDEGA

1 Introduction

Protein export by bacteria has been widely studied in *Escherichia coli* and in *Bacillus* species. The mechanism of protein translocation across the cytoplasmic membrane of these organisms appears to be comparable to the translocation of proteins across the membrane of the rough endoplasmic reticulum of eukaryotic cells, and is described in several reviews (MICHAELIS and BECKWITH 1982; SILHAVY et al. 1983; RANDALL and HARDY 1984; PUGSLEY and SCHWARTZ 1985). In gram-positive bacteria the exported proteins either remain anchored in the bacterial cell envelope, like the lipoproteins, or are released from the cells, like various proteases, amylase, and other proteins. In gram-negative bacteria the exported proteins are released into the periplasm, or are incorporated in the outer membrane of these cells. Apparently, the outer membrane of these organisms, which is both functionally and structurally completely different from the cytoplasmic membrane, forms an extra barrier for protein excretion into the extracellular space. Several gram-negative bacterial species, however, have been found to release extracellular proteins (PUGSLEY and SCHWARTZ 1985), but very little is known about the mechnisms by which proteins are translocated across the outer membrane and about the possible role of signal or other topogenic sequences in this process.

Most studies on protein export in gram-negative bacteria have been carried out with *E. coli*, although this bacterium appears to be rather restricted in its ability to excrete proteins. Well-known exceptions are hemolysins and bacter-

Department of Microbiology, Biological Laboratory, Vrije Universiteit, De Boelelaan 1087, NL-1081 HV Amsterdam

Current Topics in Microbiology and Immunology, Vol. 125

iocins. Data about the production and excretion of these proteins have indicated that *E. coli* has developed several different and specific mechanisms for their export. Export of hemolysins, for instance, is dependent on the expression of two specific gene products encoded by the hemolysin operon and involved in the translocation of the mature hemolysin across the outer membrane and its subsequent release (WAGNER et al. 1983).

In this review we describe the production and excretion of cloacin DF13 and related colicins, which are known to be synthesized as mature, soluble proteins in the cytoplasm of the producing cells. Translocation of these bacteriocins across the bacterial cell envelope and their release into the extracellular space requires the expression of a particular gene product, designated as the bacteriocin release protein (BRP) and encoded by the bacteriocinogenic plasmids. Two types of bacteriocinogenic plasmids have been described. One type, encoding the related group A colicins like colicin A, E1, E2, E3, N, and K, as well as cloacin DF13, is characterized by small (±5 kD) multi-copy-number plasmids which are not self-transmissible. The second type, encoding the related group B colicins like colicin B, Ia, Ib, and V, is characterized by large (±70 kD) low-copy-number transmissible plasmids (HARDY et al. 1973). In contrast to the group A bacteriocins, groups B bacteriocins are produced in relatively low amounts and are apparently poorly released. We describe the production and excretion of cloacin DF13 in comparison with other, mainly group A, colicins.

2 Functional Organization of Bacteriocin Operons

Analysis of the bacteriocinogenic plasmids has revealed that the region involved in bacteriocin production and release generally codes for three different genes: a gene encoding the cloacin or colicin, a gene encoding the homologous immunity protein, and a gene of which the expression is required for bacteriocin release (Fig. 1). These gene clusters can be classified into two different types on the basis of their transcriptional organization. In one type all three genes are coordinately expressed and controlled by a single inducible promoter. Transcription of the cloacin DF13 operon, for instance, is controlled by an inducible promoter upstream from the cloacin structural gene. No evident promoter structures have been detected proximal to the genes encoding the immunity protein and the bacteriocin release protein. Attenuation of transcription occurs at a ρ-dependent terminator (T1) located between the immunity gene and the BRP gene (formerly designated as gene H). A second ρ-independent and very efficient terminator (T2) is located downstream from the BRP gene (VAN DEN ELZEN et al. 1983). Potential ribosome-binding sites have been detected proximal to all three genes. These sites may provide additional possibilities to regulate the synthesis of immunity protein and bacteriocin release protein. It has been observed, for instance, that more immunity protein than cloacin protein is synthesized by pCloDF13-harboring cells (DE GRAAF and KLAASEN-BOOR 1974; KOOL et al. 1975). This difference in expression may result from differences in efficiency of the respective ribosome-binding sites. The structure of the colicin E2 operon

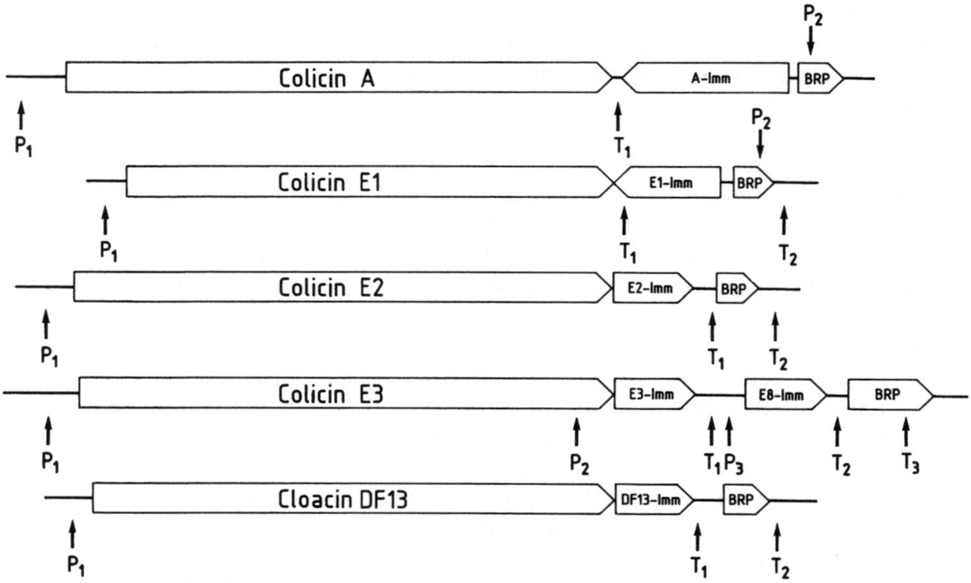

Fig. 1. Structural and functional organization of bacteriocin operons. Genes are indicated by *boxes*. The direction of transcription is indicated. P_1, inducible promoters for the synthesis of bacteriocins and bacteriocin release proteins; P_2 and P_3, promoters for immunity protein synthesis; T_1, T_2, and T_3, transcription terminators

on pColE2-P9 is similar to the cloacin DF13 operon and encodes three genes: *ceaB* for colicin E2, *ceiB* for E2-immunity protein, and *celB* for colicin release (Cole et al. 1985). Expression of all three genes is controlled by an inducible promoter upstream from the *ceaB* gene and transcription terminators have been detected downstream from the *ceiB* and *celB* gene. As shown in Fig. 1, the colicin E3 operon is essentially comparable to the cloacin DF13 and colicin E2 operons except that the ColE3-CA38 plasmid carries an additional gene coding for E8-immunity protein (Cooper and James 1984; Chak and James 1984). *E. coli* W3110 cells carrying pColE3-CA38 are immune to externally added colicin E8. Nucleotide sequence data (see below) have revealed that the structure of the E8-immunity protein is highly homologous to the sequence of the E2-immunity protein (Lau et al. 1984a, b; Watson et al. 1984; Cole et al. 1985; Masaki et al. 1985). It has been suggested that expression of the immunity genes in pColE2-P9 and pColE3-CA38 is not only dependent on the inducible colicin promoters, but also regulated by an additional promoter, independently of colicin expression (Tyler and Sherratt 1975; Mock et al. 1983; Pugsley and Schwartz 1983a; Jakes and Zinder 1984; Chak and James 1985). Nucleotide sequence data have revealed that the 3′-terminus of the colicin E2 gene (*ceaB*) contains a potential ribosome-binding site for the E2-immunity gene (*ceiB*), but an evident promoter sequence was not found (Cole et al. 1985; Masaki et al. 1985). Also, the E3- and the E2-like immunity genes of pColE3-CA38 are preceded by a ribosome-binding site (Lau et al. 1984a, b; Masaki et al. 1985). The data suggest that expression of the cloacin or colicin genes and their respective immunity genes might be translationally coupled. Translational coupling is often observed when two gene products form a protein complex like these bacteriocins.

A second type of transcriptional organization of gene clusters involved in bacteriocin production and release is found on the ColA, ColE1, and ColIb plasmids. In these clusters the colicin gene and the gene involved in colicin release form a transcriptional unit controlled by an inducible promoter, but the immunity gene which is located between these two genes is under control of a different promoter and transcribed in the opposite direction (Oka et al. 1979; Crozel et al. 1983; Sabik et al. 1983; Suit et al. 1983; Lloubès et al. 1984; Waleh and Johnson, personal communication). The stop codons of the E1-immunity gene (*imm*) and of the colicin E1 gene (*cea*) overlap with one base pair and the *imm* gene promoter is located within the coding sequence of the colicin release protein gene (*lys*, *kil*). Transcription starting at the inducible colicin E1 promoter terminates at one or two sites (T1 and T2), which have been characterized as ρ-dependent and ρ-independent terminators, respectively (Waleh and Johnson, personal communication). The stop codons for the colicin A-immunity gene (*cai*) and the colicin A gene (*caa*) are separated by a stretch of 19 base pairs (Morlon et al. 1983b; Lloubès et al. 1984). The promoter of the *cai* gene is located in the coding sequence of the gene encoding the colicin A release protein (*cal*) (Lloubès et al. 1984; Cavard et al. 1985). The organization of genes involved in production, immunity, and release of colicin N has been described. The ColN-284 plasmid shares many features with the more extensively characterized ColA and ColE1 plasmids (Pugsley 1984b).

3 Induction of Bacteriocin Synthesis

The transcription of bacteriocin operons is normally repressed by the LexA repressor protein. This protein regulates the expression of a variety of genes, in particular genes involved in the so-called "SOS" response, which are induced when *E. coli* cells are subjected to conditions that damage DNA or inhibit DNA replication (Walker 1984). Signals that activate the SOS response, like mitomycin C or UV irradiation, also induce the synthesis of bacteriocins. The inducing signal reversibly activates a specific protease, the RecA protein. The RecA protease cleaves the LexA repressor at a single site between an alanine and a glycine residue near the middle of the LexA protein (Horii et al. 1981). Intact LexA repressor protein binds at specific DNA operator sequences, containing so-called "SOS" boxes (Walker 1984). Cleavage of the bound LexA protein results in the activation of transcription of the respective operon. DNA sequence analysis of the regulatory regions of bacteriocin operons has revealed that these operons contain an SOS box (Van den Elzen et al. 1982; Ebina et al. 1983; Morlon et al. 1983a; Varley and Boulnois 1984; Cole et al. 1985; Masaki and Ohta 1985). The boxes consist in principle of an inverted repeat (Fig. 2). The complete repeat of CloDF13, and one arm of the repeat of ColA,

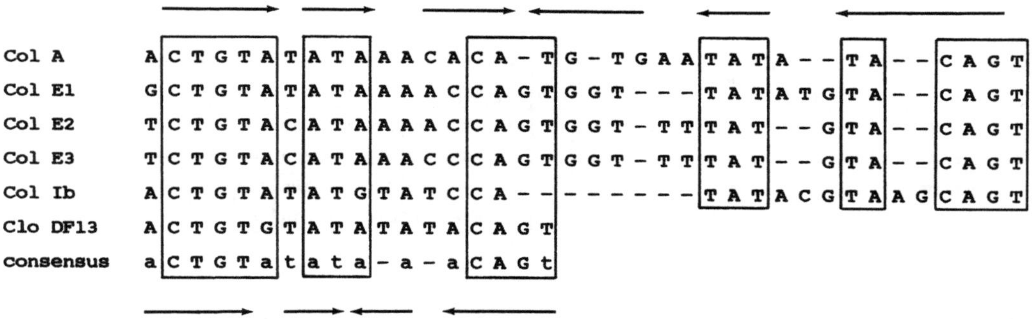

Fig. 2. Nucleotide sequences encoding the LexA-binding sites of the bacteriocin operons. *Arrows* indicate inverted repeats. Sequence homology is indicated by *boxes*. The consensus sequence for LexA-binding sites of SOS genes is given for comparison. (WALKER 1984)

ColE1, ColE2, ColE3, and ColIb, are highly homologous to the consensus sequence of the SOS box.

Induction of bacteriocin synthesis under conditions that endanger the survival of a bacterial cell or population may be envisaged as a defense mechanism against the outgrowth of related and competitive bacterial species.

4 Assignment of Functional Domains and Comparison of Primary Structures

All bacteriocins share a number of common features like the interaction with their respective immunity protein, interaction with outer membrane receptor proteins and translocation across the bacterial membranes. Furthermore, some bacteriocins exhibit a similar mode of action. Cloacin DF13 and colicin E3 inactivate bacterial ribosomes by a specific endonucleolytic cleavage of the 16S ribosomal RNA; colicin E2 and probably also E8 exhibit endonucleolytic activity on DNA, and the colicins A, E1, Ia, Ib, and K are known to form ion-permeable channels in the cytoplasmic membrane of susceptible cells (KONISKY 1982). Studies on the structure-function relationship of bacteriocins have revealed that the various functions of bacteriocin molecules can be assigned to different loci or domains. In general, the NH_2-terminal part of the bacteriocin molecules is involved in their translocation across the bacterial membranes during uptake by susceptible cells; the central part contains the receptor-binding site, and the COOH-terminal part possesses catalytic or ionophore-like activity (DE GRAAF et al. 1978; OHNO-IWASHITA and IMAHORI 1980, 1982; DANKERT et al. 1982; MOCK and PUGSLEY 1982; YAMADA et al. 1982a; BRUNDEN et al. 1984; SUIT et al. 1985). The latter domain is also involved in the interaction between certain bacteriocin molecules and their immunity proteins (DE GRAAF et al. 1978; OHNO-IWASHITA and IMAHORI 1980; MANKOVICH et al. 1984).

More indications about the location of various domains or amino acid sequences involved in comparable functions like receptor recognition, enzymatic

A

Colicin E1	Glu	Thr	Ala	Val	Ala	Tyr	Tyr	Lys	–	Asp	–	Gly	Val	Pro	Tyr	Asp	Asp	Lys	Gly	Gln
Colicin A	Pro	Gly	Phe	Asn	Tyr	Gly	Gly	Lys	Gly	Asp	–	Gly	Thr	Gly	Trp	Ser	Ser	Glu	Arg	Gly
Colicin E2	Ser	Gly	Gly	Asp	Gly	Arg	Gly	–	His	Asn	Thr	Gly	Ala	His	Ser	Thr	Ser	Gly	Asn	Ile
Colicin E3	Ser	Gly	Gly	Asp	Gly	Arg	Gly	–	His	Asn	Thr	Gly	Ala	His	Ser	Thr	Ser	Gly	Asn	Ile
Cloacin DF13	Ser	Gly	Gly	Asp	Gly	Arg	Gly	–	–	–	Pro	Gly	–	Asn	Ser	Gly	Leu	Gly	His	Asn
Colicin Ib	Ser	Asp	Pro	Val	–	Arg	Ile	–	His	Asn	Pro	Gly	Ala	Glu	Ser	Leu	Leu	Gly	Tyr	Asp

B

Imm. pr. A		Met	Asn	Glu	–	–	–	His	–	Ser	Ile	–	–	–	Asp	–	–	Thr	Asp
Imm. pr. E2				Glu	Leu	Lys	–	His	–	Ser	Ile	–	–	Ser	Asp	–	Tyr	Thr	Glu
Imm. pr. E8	Val	Lys	Val	Glu	Leu	Lys	–	–	–	Ser	Ile	–	–	Gly	Asp	–	Tyr	Thr	Glu
Imm. pr. E3			Gly	–	Leu	Lys	Leu	–	Asp	Leu	–	Thr	Trp	Phe	Asp	Lys	Ser	Thr	Glu
Imm. pr. DF13			Gly	–	Leu	Lys	Leu	His	–	–	Ile	His	Trp	Phe	Asp	Lys	Lys	Thr	Glu

Fig. 3A, B. Comparison of the NH_2-terminal amino acid sequences of various bacteriocins **A** and immunity proteins **B**. The sequences have been aligned to obtain maximum sequence homology. Deletions are indicated. Homologous amino acids are *boxed*

activity, or translocation of the proteins across the bacterial cell envelope may be obtained by a comparison of their primary structure. The NH_2-terminal amino acid sequence of cloacin DF13 and some colicins has been deduced from the nucleotide sequence data starting from the ATG translation initiation codon (Fig. 3A). The sequences have been aligned in such a way that maximal homology is obtained. Cloacin DF13 and the colicins E2 and E3 possess quite homologous NH_2-termini, but the sequence homology with other colicins appears to be restricted to a limited number of identical amino acid residues. None of the NH_2-terminal sequences possesses the characteristic features of a typical signal or leader sequence present in the precursors of most periplasmic and outer membrane proteins of *E. coli*: (1) one or more positively charged residues (Arg, Lys) at the extreme NH_2-terminus, (2) a stretch of hydrophobic or neutral residues that tends to form an α-helix, and (3) a signal peptidase cleavage site (Von Heijne 1981, 1984; Watson 1984).

The NH_2-terminal amino acid sequences of various immunity proteins show significant homology when aligned by introducing several deletions (Fig. 3B). Like the bacteriocin molecules the immunity protein molecules do not possess an NH_2-terminal leader sequence (Jakes and Model 1979; Van den Elzen et al. 1980, 1983; Varenne et al. 1981; Yamada et al. 1982b; Morlon et al. 1983b; Oudega et al. 1984b, c; Cole et al. 1985).

The absence of leader sequences has been confirmed in some cases by studies on the in vivo or in vitro translation under conditions where processing is prevented. Moreover, cloacin DF13, colicins, and immunity proteins have shown to be synthesized on free polysomes. Complexes of cloacin DF13, colicin E2, and colicin E3 with their respective immunity protein or colicins which do not form a post-translational complex with their immunity protein accumulate in the cytoplasm of producing cells, prior to their release into the periplasm and into the extracellular space (JAKES and MODEL 1979; VAN TIEL-MENKVELD et al. 1979; CAVARD et al. 1981; VARENNE et al. 1981; PUGSLEY 1983; B. Oudega unpublished results). Immunoferritin labeling of ultra-thin sections of induced, frozen cells has shown that colicin A and cloacin DF13 are exclusively located within the cell cytoplasm after induction (CAVARD et al. 1981; Lazdunski, Oudega and De Graaf, unpublished results). At no time after induction of the cells was colicin A detected in the cell periplasm or in the membranes. Analyses of the primary structure of some immunity proteins by amino acid sequencing has confirmed that these proteins are not synthesized in precursor form, but as mature proteins (VAN DEN ELZEN et al. 1980; MOCHITATE et al. 1981). Unfortunately, no such data are available for bacteriocin molecules.

The absence of an NH_2-terminal leader sequence does not exclude the possibility that functionally identical amino acid sequences are present in other parts of the molecules, but the experimental data clearly indicate that the bacteriocins are not exported by the general mechanism for protein export and translocation in *E. coli*. Recent studies on bacteriocin export mechanisms have demonstrated that the "lysis" proteins, which are responsible for the lysis and consequent death of bacteriocin-producing cells, are also involved in a more or less specific export or release of bacteriocins and immunity proteins. We suppose that other topogenic sequences, as for instance a possible interaction site for the respective bacteriocins release proteins ("lysis" proteins), are essential for the routing of bacteriocins and immunity proteins across the bacterial cell envelope. A characteristic feature of the NH_2-terminal region of cloacin DF13 and the colicins A, E1, E2, and E3, for instance, is the presence of clusters of glycine residues (not shown). Secondary structure predictions show that these regions have a high probability for β-turns (OHNO-IWASHITA and IMAHORI 1980; YAMADA et al. 1982a; MORLON et al. 1983a, b; VAN DEN ELZEN et al. 1983; COLE et al. 1985; MASAKI and OHTA 1985).

A comparison of the complete primary structure of the cloacin DF13- with the colicin E3-immunity protein, and the colicin E2- with the colicin E8-immunity protein, is presented in Fig. 4A, B. Functionally identical immunity proteins exhibit a large degree of sequence homology, but no indications for "internal" leader sequences of hydrophobic regions can be derived from the amino acid sequences.

The central regions of the various bacteriocin molecules play an essential role in their ability to bind to the respective receptor proteins located in the outer membrane of the susceptible cells. Colicin E2 and E3 share an identical amino acid sequence (res. 1 to 425) including the region which is thought to contribute to the receptor-binding activity. Both colicins recognize the same receptor protein known as the BtuB outer membrane protein, which is involved in the uptake of vitamin B12 (KONISKY 1982). Cloacin DF13 recognizes a different receptor protein involved in the uptake of ferric-aerobactin (VAN TIEL-MENKVELD et al. 1982; KRONE et al. 1983). Cloacin DF13 shows intensive sequence homology with the colicins E2 and E3 in the region of residues 100–340 in the cloacin sequence, but the region of residues 340–430 is not homologous. We propose that this region, corresponding to the region of residues 330–420 in colicin E2 and E3, specifies the receptor-binding activity.

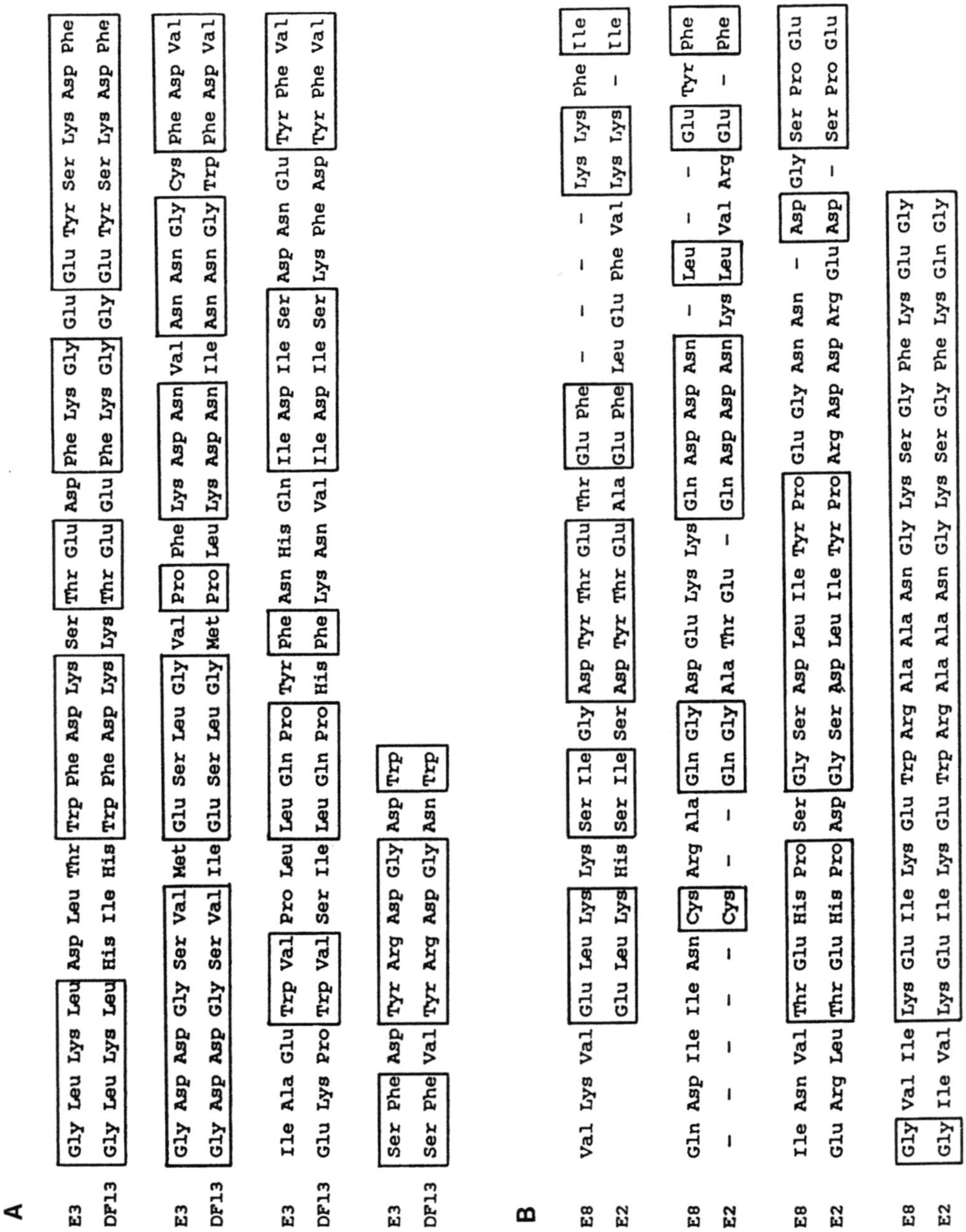

Fig. 4A, B. Homology in the primary structure of colicin E3- and cloacin DF13-immunity protein **A** and in colicin E8- and colicin E2-immunity protein **B**

			431																447						
DF13	Arg	Asp	Lys	Ala	Ala	Leu	Ser	Gly	Ala	Met	Glu	Ser	Arg	Lys	Gln	Lys	Glu	Lys	Lys	Ala	Lys	Glu	Ala	Glu	Ala
E3	Asp	Ala	Asp	Ala	Ala	Leu	Ser	Ser	Ala	Met	Glu	Ser	Arg	Lys	Lys	Lys	Glu	Asp	Lys	Lys	Arg	Ser	Ala	Glu	–
E2	Asp	Ala	Asp	Ala	Ala	Leu	Ser	Ala	Ala	Gln	Glu	Arg	Arg	Lys	Gln	Lys	Glu	Asn	Lys	Glu	Lys	Asp	Ala	Lys	–
			421																437						

DF13	Glu	Asn	Lys	Leu	Asn	Glu	Glu	–	Lys	Lys	–	Lys	Pro	Arg	Lys	–	–	–	–	–	–	–	–	Gly	Thr
E3	Asn	Asn	–	Leu	Asn	Asp	Glu	–	Lys	–	Asn	Lys	Pro	Arg	Lys	–	–	–	–	–	–	–	–	Gly	Phe
E2	Asp	–	Lys	Leu	Asp	Lys	Glu	Ser	Lys	Arg	Asn	Lys	Pro	Gly	Lys	Ala	Thr	Gly	Lys	Gly	Lys	Pro	Val	Gly	Asp

DF13	Lys	–	–	–	–	–	–	–	–	–	–	–	–	–	–	–	–	–	–	Asp	Tyr	Gly	His	Asp	Tyr
E3	Lys	–	–	–	–	–	–	–	–	–	–	–	–	–	–	–	–	–	–	Asp	Tyr	Gly	His	Asp	Tyr
E2	Lys	Trp	Leu	Asp	Asp	Ala	Gly	Lys	Asp	Ser	Gly	Ala	Pro	Ile	Pro	Asp	Arg	Ile	Ala	Asp	Lys	Leu	Arg	Asp	Lys

		475														489									
DF13	Phe	Pro	Asp	Pro	Lys	Thr	Glu	Asp	Ile	Lys	Gly	Leu	Gly	Glu	Leu	Lys	Glu	Gly	Lys	Pro	Lys	Thr	Pro	Lys	Gln
E3	His	Pro	Ala	Pro	Lys	Thr	Glu	Asn	Ile	Lys	Gly	Leu	Gly	Asp	Leu	Lys	Pro	Gly	Ile	Pro	Lys	Thr	Pro	Lys	Gln
E2	Glu	Phe	Lys	Asn	Phe	Asp	Asp	Phe	Arg	Lys	Lys	Leu	Trp	Glu	Glu	Val	Ser	Lys	Asp	Pro	Asp	Leu	Ser	Lys	Gln

DF13	Gly	Gly	Gly	Gly	Lys	Arg	Ala	Arg	Trp	Tyr	Gly	Asp	Lys	Lys	Arg	Lys	Ile	Tyr	Glu	Trp	Asp	Ser	Gln	His	Gly
E3	Asn	Gly	Gly	Gly	Lys	Arg	Lys	Arg	Trp	Thr	Gly	Asp	Lys	Gly	Arg	Lys	Ile	Tyr	Glu	Trp	Asp	Ser	Gln	His	Gly
E2	Phe	Lys	Gly	Ser	Asn	Lys	Thr	Asn	Ile	Gln	Lys	Gly	Lys	Ala	Pro	Phe	Ala	Arg	Lys	Lys	Asp	Gln	Val	Gly	Gly

DF13	Glu	Leu	Glu	Gly	Tyr	Arg	Ala	Ser	Asp	Gly	Glu	His	Leu	Gly	Ala	Phe	Asp	Pro	Lys	Thr	Gly	Lys	Gln	Val	Lys
E3	Glu	Leu	Glu	Gly	Tyr	Arg	Ala	Ser	Asp	Gly	Gln	His	Leu	Gly	Ser	Phe	Asp	Pro	Lys	Thr	Gly	Asn	Gln	Leu	Lys
E2	Arg	Glu	Arg	Phe	Glu	Leu	His	His	Asp	Lys	Pro	Ile	Ser	Gln	Asp	Gly	Gly	Val	Tyr	Asp	Met	Asn	Asn	Ile	Arg

												560		
DF13	Gly	Pro	Asp	Pro	Lys	Arg	Asn	Ile	Lys	Lys	Tyr	Leu		
E3	Gly	Pro	Asp	Pro	Lys	Arg	Asn	Ile	Lys	Lys	Tyr	Leu		
E2	Val	Thr	Thr	Pro	Lys	Arg	His	Ile	Asp	Ile	His	Arg	Gly	Lys
														580

Fig. 5. Comparison of the COOH-terminal amino acid sequence of cloacin DF13, colicin E3, and colicin E2. Identical amino acid sequences are *boxed*

Remarkably, the colicins E1 and A, which use the same BtuB receptors as colicin E2 and E3, show no significant sequence homology with the latter colicins (MOCK and PUGSLEY 1982). On the other hand, colicin A and E1 also share a common amino acid sequence in their receptor-binding domain (CAVARD and LAZDUNSKI 1981; MORLON et al. 1983b). A very similar amino acid sequence is found in colicin Ib, which uses the *cir* gene product for its uptake by susceptible cells (VARLEY and BOULNOIS 1984). In contrast to colicin E2 and E3, the receptor for colicin A is a complex of the OmpF protein, the *BtuB* gene product, and lipopolysaccharide (CAVARD and LAZDUNSKI 1981; CHAI et al. 1982). Possibly, the colicins A, E1, and Ib, which have a similar mode of action on the cytoplasmic membrane, require other outer membrane components for their uptake in addition to the *BtuB* of *cir* gene products.

Figure 5 shows a comparison of the COOH-terminal amino acid sequences of cloacin DF13, colicin E3, and colicin E2. These protein fragments contain the nuclease activity and the interaction site with the respective immunity proteins (DE GRAAF et al. 1978; OHNO-IWASHITA and IMAHORI 1980; MASAKI and OHTA 1982; MOCK et al. 1983; COLE et al. 1985). Cloacin DF13 and colicin E3 possess similar endoribonucleolytic activity, and the catalytic domains in these bacteriocins show almost complete sequence homology. The sequence homology with colicin E2 remains restricted to a few amino acid residues, in particular at the extreme COOH-terminal part of the molecules. In between the amino acid sequences known to be (part of) the receptor-binding and the catalytic domain, a region of significant sequence homology is observed for all three bacteriocins. A comparable region is also found in colicin A and E1 (Fig. 6A). It is not known whether this region is part of one of the two domains mentioned above or constitutes another common feature of all these bacteriocins. In this respect it should be noted that a remarkable and significant sequence homology is found between cloacin DF13 of colicin E3 and their respective immunity proteins at the beginning of the catalytic domains (Fig. 6B). It remains to be investigated whether these sequences are involved in the specific interaction of both proteins with their BRPs, or whether these sequences play an essential role in the translocation across the cell envelope during their uptake by susceptible cells.

Contrary to the bacteriocins and immunity proteins, the genes responsible for bacteriocin release and "lysis" of the producing cells at certain conditions encode for proteins that are synthesized with a typical leader sequence which might be recognized by the lipoprotein signal peptidase (OKA et al. 1979; HAKKAART et al. 1981a; NIELSEN and LAMPEN 1982; OUDEGA et al. 1982, 1984c; CAVARD et al. 1985; CHAK and JAMES 1985; COLE et al. 1985; Waleh and Johnson, personal communication).

The mature gene products show a very high degree of sequence homology (Fig. 7). The type of the predicted cleavage site suggests that these bacteriocin release proteins, like the lipoproteins, may be subjected to post-translational modification by substitution with a diglyceride moiety in a thioether linkage and fatty acyl groups at the NH_2-terminal cysteine residue (INUKAI et al. 1984).

In agreement with this assumption is the observation that the mature pColE2-bacteriocin release protein (*celB* gene product) is not detectable in cells

A

Cloacin DF13	Asp	Ala	Leu	Ser	Arg	Asp	Lys	Ala	Ala	Leu	Ser	Gly	Ala	Met	Glu	Ser	Arg	Lys	Gln	Lys	Glu	Lys	Lys	(res. 425–447)
Colicin E3	Lys	Glu	Lys	Ser	Asp	Ala	Asp	Ala	Ala	Leu	Ser	Ser	Ala	Met	Glu	Ser	Arg	Lys	Lys	Lys	Glu	Asp	Lys	(res. 415–437)
Colicin E2	Lys	Glu	Lys	Ser	Asp	Ala	Asp	Ala	Ala	Leu	Ser	Ala	Ala	Gln	Glu	Arg	Arg	Lys	Gln	Lys	Glu	Asn	Lys	(res. 415–437)
Colicin A	Asn	Glu	Lys	Ser	Ala	Ala	Asp	Ala	Ala	Leu	Ser	Ala	Ala	Thr	Ala	Gln	Arg	Leu	Gln	Ala	Glu	Ala	Ala	(res. 338–360)
Colicin E1	Leu	Glu	Lys	Lys	Ala	Ala	Asp	Ala	Gly	Val	Ser	Tyr	Val	Val	Ala	Leu	Leu	Phe	Ser	Leu	Leu	Ala	Gly	(res. 466–488)

B

DF13–Imm. pr.	Phe	Asp	Lys	Lys	–	Thr	Glu	Glu	Phe	Lys	Gly	–	Gly	Glu	Tyr	Ser	Lys	(res. 9–23)
Cloacin DF13	Pro	Asp	Pro	Lys	–	Thr	Glu	Asp	Ile	Lys	Gly	Leu	Gly	Glu	Leu	–	Lys	(res.475–489)
Colicin E3	Pro	Ala	Pro	Lys	–	Thr	Glu	Asn	Ile	Lys	Gly	Leu	Gly	Asp	Leu	–	Lys	(res.465–479)
E3–imm. pr.	Phe	Asp	–	Lys	Ser	Thr	GLu	Asp	Phe	Lys	Gly	–	Glu	Glu	Tyr	Ser	Lys	(res. 9–23)

Fig. 6A, B. Homologous amino acid sequences in the central region of the primary structure of various bacteriocins **A** and in the primary structure of cloacin DF13, colicin E3, and their respective immunity proteins **B**

DF13	Met	Lys	Lys	Ala	Lys	Ala	Ile	Phe	Leu	Phe	Ile	Leu	Ile	Val	Ser	Gly	Phe	Leu	Leu	Val	Ala	Cys	Gln	Ala	Asn
E1	Met	Arg	Lys	–	Lys	Phe	–	Phe	Val	Gly	Ile	Phe	Ala	–	Ile	–	Asn	Leu	Leu	Val	Gly	Cys	Gln	Ala	Asn
E2	Met	Lys	Lys	Ile	Thr	Gly	Ile	Ile	Leu	Leu	Leu	Leu	Ala	Val	Ile	–	–	Ile	Leu	Ser	Ala	Cys	Gln	Ala	Asn
E3	Met	Lys	Lys	Ile	Thr	Gly	Ile	Ile	Leu	Leu	Leu	Leu	Ala	Val	Ile	–	–	Ile	Leu	Ser	Ala	Cys	Gln	Ala	Asn
A	Met	Lys	Lys	Ile	–	Ile	Ile	Cys	Val	Ile	Leu	Leu	Ala	–	Ile	–	Met	Leu	Leu	Ala	Ala	Cys	Gln	Val	Asn

DF13	Tyr	Ile	Arg	Asp	Val	Gln	Gly	Gly	Thr	Val	Ala	Pro	Ser	Ser	Ser	Ser	Glu	Leu	Thr	Gly	Ile	Ala	Val	Gln	
E1	Tyr	Ile	Arg	Asp	Val	Gln	Gly	Gly	Thr	Ile	Ala	Pro	Ser	Ser	Ser	Ser	Lys	Leu	Thr	Gly	Ile	Ala	Val	Gln	
E2	Tyr	Ile	Arg	Asp	Val	Gln	Gly	Gly	Thr	Val	Ser	Pro	Ser	Ser	Thr	Ala	Glu	Val	Thr	Gly	Leu	Ala	Thr	Gln	
E3	Tyr	Ile	Arg	Asp	Val	Gln	Gly	Gly	Thr	Val	Ser	Pro	Ser	Ser	Thr	Ala	Glu	Val	Thr	Gly	Leu	Ala	Thr	Ser	– 50 res.
A	Asn	Val	Arg	Asp	Thr	Gly	Gly	Gly	Ser	Val	Ser	Pro	Ser	Ser	Ile	Val	Thr	Gly	Val	Ser	Met	Gly	Ser	Asp	– 5 res.

Fig. 7. The primary structure of bacteriocin release proteins including their NH_2-terminal signal sequence. Homologous sequences are *boxed*. The probable cleavage site for the signal peptidase is indicated by an *arrow*. The colicin E3-release protein is supposed to be much larger than the other bacteriocin release proteins, because of the apparent deletion of one cytosine residue in between the two last codons of the nucleotide sequence encoding the other bacteriocin release proteins (Watson et al. 1984)

treated with globomycin which is a specific inhibitor of the lipoprotein signal peptidase (COLE et al. 1985). Processing of the pColA-encoded bacteriocin release protein (the *cal* gene product) is also inhibited by globomycin, but the processing of the pCloDF13-bacteriocin release protein appeared not to be hampered by globomycin (Lazdunski, personal communication; OUDEGA et al. 1984c).

5 Function(s) of Bacteriocin Release Proteins

Cloacin DF13, colicin E2, and colicin E3 form strong equimolar complexes with their respective immunity proteins inside the cytoplasm of the producing cells, and these bacteriocins are subsequently released from the cells as a complex of both polypeptides (JAKES and ZINDER 1974; SCHALLER and NOMURA 1976; DE GRAAF and KLAASEN-BOOR 1977).

In contrast, the colicins A, E1, Ia and Ib do not form a complex with their immunity protein, but leave the producing cell as single polypeptides (KONINSKY 1982). In early reports on the mechanism of colicin release it was claimed that following induction of colicin synthesis the bacteriocin complexes first accumulate in the cytoplasm and subsequently leave the cells by a nonspecific mechanism called actual lysis, or apparent-, semi-, or quasi-lysis, resulting in an increased permeability of the cell envelope of the producing cells. The simultaneous release of intracellular enzymes, including the elongation factor G and Tu, and β-galactosidase, and the release of the periplasmic acid and alkaline phosphatase, was considered as further evidence for a nonspecific release mechanism (MOCK and SCHWARTZ 1978; PUGSLEY and ROSENBUSCH 1981; VARENNE et al. 1981; PUGSLEY 1983; JAKES and ZINDER 1984; PUGSLEY and SCHWARTZ 1984). Furthermore, it was observed that within 1 h after induction of cells carrying pColE1, macromolecular synthesis stops, the capacity for transport of proline, thiomethyl-β-D-galactoside, and α-methyl glucoside is lost, and the cytoplasmic membrane becomes abnormally permeable to *o*-nitrophenyl-β-D-galactoside (SUIT et al. 1983). In these studies, however, the producing cells were fully induced with relatively high amounts of mitomycin C, which has been shown to cause a decline in culture turbidity, cell lysis, and death. More detailed studies about the effect of variable concentrations of mitomycin C on the production and release of cloacin DF13 have demonstrated that the expression of the cloacin DF13 operon can be regulated by varying the mitomycin C concentration (VAN TIEL-MENKVELD et al. 1979, 1981; OUDEGA et al. 1982). An almost optimal production of cloacin DF13, without significant lysis of the cells, is obtained when the cells are treated with relatively low mitomycin C concentrations. Under these conditions the cloacin DF13 complexes first accumulate in the cytoplasm, thereafter in the periplasm, and finally in the extracellular space. More than half of the bacteriocin molecules (about 50–100 mg/liter) are released within a few hours after induction. Furthermore, the producing cells do not release significant amounts of cytoplasmic β-galactosidase or periplasmic β-lactamase under these conditions. Apparently, treatment of bacteriocinogenic cells with low concentrations of mitomycin C is sufficient to induce both bacteriocin synthesis and release, while higher concentrations have an additional and marked effect on membrane permeability, resulting in cell lysis. Treatment of nonbacteriocinogenic cells with comparable high mitomycin C concentrations does not induce lysis, indicating that the presence of bacteriocinogenic plasmids is required for the observed phenomenon. Recently, it has been shown that the mitomycin C-induced lysis is strain dependent. Some strains release bacteriocin without apparent lysis even when fully induced (Oudega, unpublished results; A.P. Pugsley, personal communication).

Analysis of various pCloDF13 insertion and deletion mutants has shown that transcription of pCloDF13-DNA sequences adjacent to the cloacin gene is essential for mitomycin C-induced lethality and cloacin DF13 release (Van Tiel-Menkveld et al. 1981; Oudega et al. 1982). A small gene, denominated gene H, and located downstream of the CloDF13 immunity gene, has been identified and found to be responsible for these processes (Hakkaart et al. 1981 a, b). Studies with mutants impaired in the synthesis of protein H encoded by this gene have indicated that protein H is essential for the release of cloacin DF13 (Oudega et al. 1982). *E. coli* cells harboring mutant CloDF13 plasmids with a deletion, or with a transposon insertion in gene H, accumulate cloacin DF13 in their cytoplasm. Comparable genes involved in the lysis of colicinogenic cells have been identified (Schafferman et al. 1979; Pugsley and Schwartz 1983a, b; Jakes and Zinder 1984; Yamada and Nakazawa 1984; Cavard et al. 1985). The pColE1 *kil* gene product, the pColE2-P9 *celB* gene product, the pColE3-CA38 *lys* gene product, and the pColA-CA31 *cal* gene product are also essential for the release of their respective colicins. The *ceiB* gene encoding colicin E2-immunity protein and the *celB* gene were cloned under control of the *lac* promoter (Pugsley and Schwartz 1983b). Experiments with cells carrying these cloned genes revealed that the expression of *celB* is sufficient to cause cell lysis in the absence of colicin E2 production. Recent experiments on the induction of colicin E2 synthesis with variable mitomycin-C concentrations have supported the hypothesis that colicin E2 is released by a "semiselective" increase in membrane permeability, allowing substantial amounts of colicin E2, but only minor amounts of other proteins, to be released from the cells (Pugsley and Schwartz 1984).

The mechanism by which the products of the "lysis" genes induce cell lysis when synthesized in high amounts, or facilitate more or less specific bacteriocin release when synthesized in relatively low amounts, is still largely unknown. It seems rather obvious, however, that the primary role of these proteins is to release the intracellular bacteriocin from the cell. For this reason we propose to call these proteins bacteriocin release proteins (BRP). We assume that induction with low mitomycin C concentrations of cells carrying a multi-copy plasmid results in an intermediate synthesis of BRP, because not all DNA-bound LexA repressor molecules become inactivited. This intermediate amount of BRP, however, is sufficient to facilitate a more or less specific release of bacteriocin molecules into the extracellular space. Full induction of all bacteriocin operons present in the cell gives rise to high levels of BRP, ultimately resulting in lysis of the producing cell.

The mature BRPs share a substantial structural homology and appeared to be functionally interchangeable with respect to their ability to promote the release of heterologous bacteriocins (Pugsley and Schwartz 1983a).

With the exception of the pColA-encoded release protein, all mature BRPs have been localized in the cell envelope of the induced cells (Hakkaart et al. 1981a; Jakes and Zinder 1984; Cole et al. 1985). Furthermore the cloacin DF13 release protein has been detected in the cytoplasmic and outer membrane (Oudega et al. 1984c). The colicin A release protein may be excreted into the medium together with colicin A (Cavard et al. 1985). As mentioned before the processing of the colicin E2 pro-release protein is inhibited by globomycin, a specific inhibitor of the lipoprotein signal peptidase. These observations may indicate that this release protein and the lipoprotein might have comparable features (Cole et al. 1985). Post-translational modification at the NH_2-terminal cysteine residue, however, has not been detected, and the mature release proteins possess no significant sequence homology with the mature lipoprotein except for a conserved Asn, Asp, and Val residue at position 4, 8, and 13, respectively (Braun 1975). Similar results have been obtained with the cloacin DF13 release protein, but the processing of this protein could not be inhibited with globomycin (Oudega et al. 1984c).

6 Effects of Culture Conditions on Bacteriocin Release and the Role of Phospholipase A in this Process

Growth in the presence of fermentable sugars hampers the synthesis and release of cloacin DF13 (Van Tiel-Menkveld et al. 1979). About a tenfold higher concentration of mitomycin C is required to induce the same level of cloacin DF13 production as observed in a culture grown with lactate as carbon source (Van Tiel-Menkveld et al. 1981). Addition of cAMP restores the synthesis, but not the release of the bacteriocin, indicating that the inhibitory effect of fermentable sugars on production is mediated by catabolite repression (Oudega et al. 1982; Van den Elzen et al. 1982). Comparable results have been described for colicin E1 and Ib (Nakazawa and Tamada 1972; Durkacz et al. 1974; Isaacson and Konisky 1974; Pugsley 1981). Analysis of the regulatory region of the respective operons has revealed the presence of a potential cAMP-CRP-binding site (Van den Elzen et al. 1982; Varley and Boulnois 1984). Further studies on the inhibiting effect of fermentable sugars on cloacin release have demonstrated that this effect is caused by a rather strong acidification of the culture medium due to the accumulation of acidic fermentation products such as formate, acetate, and lactate (Oudega et al. 1984b). These weak organic acids have an uncoupling effect on oxidative phosphorylation when they accumulate during fermentation reactions (Hueting and Tempest 1977; De Vries et al. 1980; Linton et al. 1981). The subsequent deenergization of the cytoplasmic membrane results in a strong inhibition of processing and accumulation of nonfunctional cloacin DF13 release protein (Oudega et al. 1984b).

Addition of divalent cations, such as Mg^{2+} and Ca^{2+}, protects cloacinogenic cells against the BRP-induced lysis observed after induction with relatively high concentrations of mitomycin C, but does not affect the synthesis or release of the bacteriocin (OUDEGA et al. 1984a; LUIRINK et al. 1985). The optimal concentration of Mg^{2+} or Ca^{2+} ions is 20 mM and 10 mM, respectively. Similar results have been described for colicin E2 (PUGSLEY and SCHWARTZ 1984). Induction of colicin E2 synthesis with suboptimal mitomycin C concentrations, or full induction in the presence of 20 mM Mg^{2+}, results in colicin E2 release without significant lysis of the producing cells. Under these conditions colicin E2 was by far the major protein detected in the culture medium, while other proteins could only be detected in very small amounts. The experimental data clearly suggest that bacteriocin release is not necessarily coupled with cell lysis, and that the bacteriocin release is not the event causing the observed decline in culture turbidity. Under the proper conditions more than 90% of the cloacin DF13 produced is released from the cells before any decline in culture turbidity or degradation of peptidoglycan can be measured (LUIRINK et al. 1985).

Addition of Triton X-100 has a positive effect on the release of colicin E2, colicin M, and cloacin DF13 (PUGSLEY and SCHWARTZ 1984; LUIRINK et al. 1985; V. Braun, personal communication). Also high concentrations of salt provoke the release of colicins (HERSCHMAN and HELINSKI 1967; MOCK and SCHWARTZ 1978; PUGSLEY and SCHWARTZ 1984).

The first indications about the mechanism of BRP-induced cell lysis have been obtained recently. Expression of the pColE2-release protein renders the cells extremely sensitive to lysozyme, which demonstrates that the permeability of the outer membrane has been altered (PUGSLEY and SCHWARTZ 1984). This change in permeability could be attributed, in part, to the colicin E2-release protein-dependent activation of the detergent-resistant phospholipase A, present in the outer membrane. It was proposed that lysophosphatidylethanolamine, the product of the action of phospholipase A on phosphatidylethanolamine, the major outer membrane phospholipid, might act as a membrane perturbant which could alter the permeability properties of the envelope and allow colicin E2 to leak out of the cell.

Comparable results were obtained with cloacin DF13 (OUDEGA et al. 1984a; LUIRINK et al. 1985). Cloacinogenic cells defective in both detergent-resistant and detergent-sensitive phospholipase A were unable to release cloacin DF13, but considerable amounts of this bacteriocin accumulated in the periplasm and in the cytoplasm. Complementation of the mutation affecting detergent-resistant phospholipase A activity by introducing a plasmid encoding this phospholipase A restored normal excretion of cloacin DF13. This indicates that detergent-resistant phospholipase A activity in the outer membrane plays a major role in the BRP-dependent release of cloacin DF13.

7 Release of Mutant Bacteriocin Molecules and Bacteriocin Fragments

The more or less specific or semispecific nature of bacteriocin release at the proper culture conditions might be due to the fact that bacteriocins have the appropriate properties to be released from the cells through the BRP-induced permeabilization zones in the membranes. Another explanation, however, might be that bacteriocins may have specific, so-called topogenic sequences or tertiary structures, which are essential for efficient recognition of, and export through, the permeabilization zones. The COOH-terminal portion of the colicin E1 molecule is necessary for efficient export especially through the cytoplasmic membrane (YAMADA and NAKAZAWA 1984) and might contain an internal signal-like sequence, which plays a role in this process. Mutant colicin E3 molecules have been described that have the same apparent molecular weight as wild-type colicin E3, but failed to be exported to the cell surface (MOCK and SCHWARTZ 1978). Export of these mutant molecules appeared to be blocked at some stage in the export process. Moreover, the mutant proteins were also unable to penetrate the outer membrane of colicin E3-susceptible cells.

Stable, truncated polypeptides comprising the NH_2-terminal half of colicin E2 or the NH_2-terminal 75% of colicin N are exported as efficiently as the full-length colicins E2 and N (PUGSLEY 1984a). As mentioned before, the NH_2-terminal regions of several bacteriocins, which contain essential information necessary for uptake (KONISKY 1982; DAVIDSON et al. 1984), are surprisingly rich in glycine residues and show structural homology in the first part of the NH_2-terminal region.

Three types of CloDF13 mutant plasmids have been used to study the export of mutant cloacin DF13 molecules and of cloacin fragments (OUDEGA et al. 1984a). By chemical mutagenesis a pCloDF13 mutant with a point mutation in the cloacin gene has been isolated (ANDREOLI and NIJKAMP 1976). This mutation has been identified and the mutant protein contains a serine residue instead of a proline residue at position 52 (Luirink et al. unpublished results). Cells harboring this mutant CloDF13 plasmid, designated pCloDF13-clpO3, produce normal amounts of mutant cloacin DF13 molecules upon induction, but the molecules possess no killing activity, due to a hampered translocation across the outer membrane of susceptible cells (GAASTRA et al. 1980). In contrast to the colicin E3 mutants (MOCK and SCHWARTZ 1978) the mutant cloacin DF13 molecules are excreted as wild-type cloacin DF13.

CloDF13 mutant plasmids with an insertion of transposon Tn*901* at different positions in the cloacin gene (ANDREOLI et al. 1978) were used to study the export of truncated cloacin molecules shortened at their COOH-terminus. These cloacin fragments accumulate in the cell cytoplasm and no lysis of host cells is observed after strong induction. Obviously, the cells do not synthesize the cloacin release protein due to a polar effect of the transposon insertion on the expression of gene H.

Cells harboring the deletion mutant pVU215 (HAKKAART et al. 1981a) synthesize an NH_2-terminal cloacin fragment of 42 amino acid residues (molecular weight about 5000). Using an enzyme-linked-immunosorbent assay this fragment was found to be excreted into the extracellular space upon induction of the cells. The same fragment accumulates in the cytoplasm of cells harboring the mutant plasmid pVU211, which no longer encodes for the release protein (HAKKAART et al. 1981a). The data suggest that the NH_2-terminal region of cloacin might contain essential information required for export.

Recently it has been found by using monoclonal antibodies that the cloacin DF13 immunity protein was also produced and excreted in the absence of cloacin molecules (Luirink et al., unpublished results).

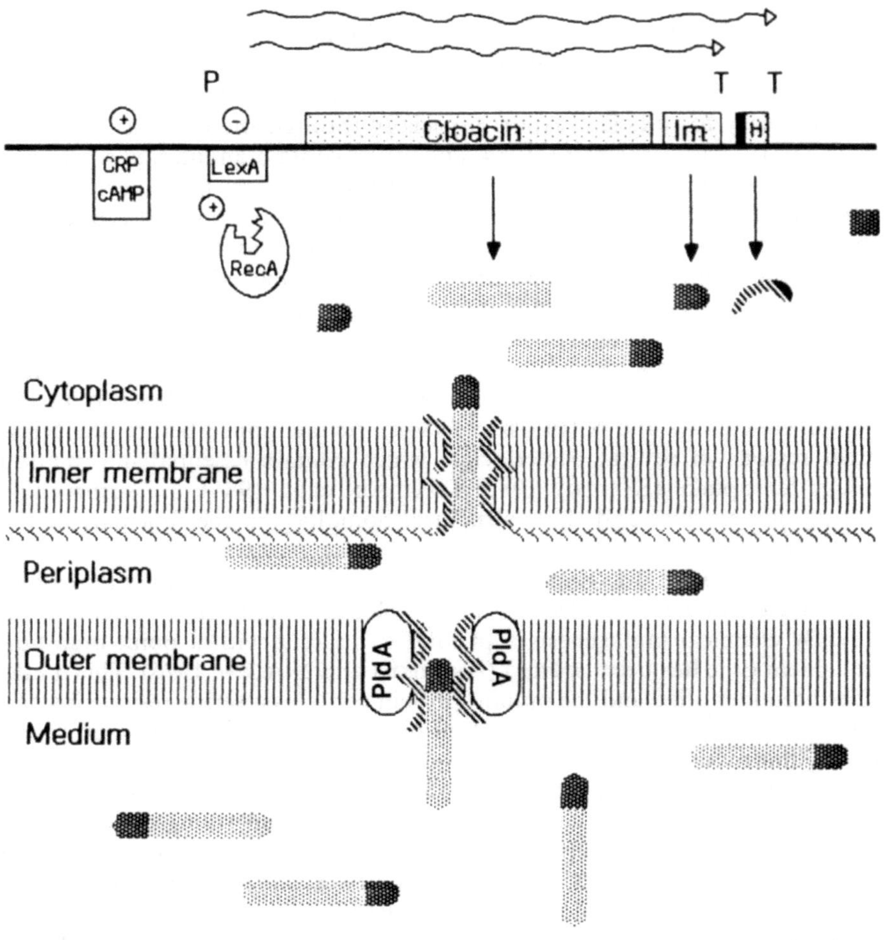

Fig. 8. Model for the production and release of cloacin DF13. *P*=inducible cloacin promoter; *T*=transcription terminators; *Im*=cloacin DF13-immunity protein gene; *H*=cloacin DF13-release protein gene; *Pld A*=detergent-resistent phospholipase A

8 Model and Possible Future Applications

A simplified model for the production and release of cloacin DF13 has been developed (Fig. 8). Upon derepression of the promoter, cloacin and immunity protein are synthesized in high amounts and the two polypeptides form an equimolar complex in the cell cytoplasm. An excess of immunity protein provides the cell with cytoplasmic immunity against cloacin molecules. Bacteriocin release proteins are synthesized with an NH_2-terminal signal sequence. The precursors of these proteins are translocated and inserted into both the cytoplasmic and outer membrane. Following processing and possibly modification of the NH_2-terminal cysteine residue of the mature bacteriocin release proteins, these molecules form some kind of pores or permeabilization zones in both membranes. In the outer membrane, detergent-resistant phospholipase A is activated, either directly by the bacteriocin release protein or indirectly as a result of the "pore"-forming ability of this protein. Cloacin DF13 complexes accumulate in the cell cytoplasm, recognize the permeabilization zones, and traverse both membranes through these "pores." The presence of the bacteriocin release protein in the cytoplasmic membrane is essential for this translocation process. Activation of detergent-resistant phospholipase A is essential for an efficient release of the bacteriocin across the outer membrane. The amino terminal part of the cloacin and possibly also of the immunity protein molecule might contain structural information required for the recognition of the permeabilization zones. Besides cloacin DF13 complexes, immunity protein and minor amounts of other soluble proteins might be released through the permeabilization zones. A high number of these permeabilization zones, and thus a high phospholipase A activity, results in severe membrane damage and in activation of the autolytic system of *E. coli* (LEDUC et al. 1982). This activation of the autolytic system ultimately results in cell lysis and death. Divalent cations such as Mg^{2+} and Ca^{2+} hamper autolysis in *E. coli* as well as BRP-induced lysis, probably by stabilizing the bacterial membranes. The supposed pore-forming ability of the bacteriocin release proteins is not a unique phenomenon, because other small polypeptides such as, for instance, melittin and the S gene product of phage lambda, are known to produce pores in lipid bilayer membranes (TERWILLIGER and EISENBERG 1982; ALTMAN et al. 1983). Moreover, the melittin-induced membrane disruption is thought to enhance the activity of phospholipase A2.

The application of bacteriocin release systems for export of other (foreign) proteins might be advantageous for the production and isolation of commercially interesting proteins such as, for instance, vaccine antigens, hormones, and other polypeptides. In *E. coli*, gene fusions have been made to promote the production and export of foreign proteins to the periplasm, by fusing the gene encoding the foreign protein to the 3′- end of a bacterial gene encoding the leader sequence of an exported protein (TALMADGE et al. 1980; SUZUKI et al. 1982; TALMADGE and GILBERT 1982; OHMURA et al. 1984). However, instability of the hybrid gene, as well as poor secretion and proteolytic degradation of the hybrid gene products, have complicated successful applications. The bacteriocin release system is an unique alternative, because soluble cytoplasmic proteins are exported in their native conformation, without the "interference" of a signal sequence and processing. Fusions between the structural gene encoding a foreign protein and the 3′- end of a bacteriocin gene encoding the NH_2-terminal, glycine-rich part of this molecule might result in the expression of hybrid gene products which might be released by producing cells as wild-type bacteriocins. Such fusions are presently constructed and a suitable host-vector system for the production and release of soluble foreign proteins by *E. coli* is being developed.

References

Altman E, Altman RK, Garrett JM, Grimaila RJ, Young R (1983) *S* gene product: identification and membrane localization of a lysis control protein. J Bacteriol 155:1130–1137

Andreoli PM, Nijkamp HJJ (1976) Mutants of the CloDF13 plasmid in *Escherichia coli* with a decreased bacteriocinogenic activity. MGG 144:159–170

Andreoli PM, Overbeeke N, Veltkamp E, Van Embden JDA, Nijkamp HJJ (1978) Genetic map of the bacteriocinogenic plasmid CloDF13 derived by insertion of the transposon Tn*901*. MGG 160:1–11

Braun V (1975) Covalent lipoprotein from the outer membrane of *Escherichia coli*. Biochim Biophys Acta 415:335–377

Brunden KR, Cramer WA, Cohen FS (1984) Purification of a small receptor-binding peptide from the central region of the colicin E1 molecule. J Biol Chem 259:190–196

Cavard D, Lazdunski C (1981) Involvement of BtuB and OmpF proteins in binding and uptake of colicin A. FEMS Microbiol Lett 12:311–316

Cavard D, Bernadac A, Lazdunski C (1981) Exclusive localization of colicin A in cell cytoplasm of producing bacteria. Eur J Biochem 119:125–131

Cavard D, Lloubès R, Morlon J, Chartier M, Lazdunski C (1985) Lysis protein encoded by plasmid ColA-CA31. Gene sequence and export. MGG 199:95–100

Chai T-J, Wu V, Foulds J (1982) Colicin A receptor protein: role of two *Escherichia coli* outer membrane proteins (OmpF protein and *BtuB* gene product) and lipopolysaccharide. J Bacteriol 151:983–988

Chak K-F, James R (1984) Localization and characterization of a gene on the ColE3-CA38 plasmid that confers immunity to colicin E8. J Gen Microbiol 139:701–710

Chak K-F, James R (1985) Analysis of the promoters for the two immunity genes present in the ColE3-CA38 plasmid using two new promoter probe vectors. Nucleic Acids Res 13:2519–2531

Cole ST, Saint-Joanis B, Pugsley AP (1985) Molecular characterization of the colicin E2 operon and identification of its products. MGG 198:465–472

Cooper PC, James R (1984) Two new E colicins, E8 and E9, produced by a strain of *Escherichia coli*. J Gen Microbiol 130:209–215

Crozel V, Lazdunski C, Cavard D (1983) Localization of genes responsible for replication and immunity to colicin A on plasmid ColA-CA31. MGG 192:500–505

Dankert JR, Uratani Y, Grabau C, Cramer WA, Hermodson M (1982) On a domain structure of colicin E1. J Biol Chem 257:3857–3863

Davidson VL, Brunden KR, Cramer WA, Cohen FS (1984) Studies on the mechanism of action of channel-forming colicins using artificial membranes. J Membr Biol 79:105–118

De Graaf FK, Klaasen-Boor P (1974) Purification and characterization of the cloacin DF13 immunity protein. FEBS Lett 40:293–296

De Graaf FK, Klaasen-Boor P (1977) Purification and characterization of a complex between cloacin and its immunity protein isolated from *Enterobacter cloacae* (CloDF13). Eur J Biochem 173:107–114

De Graaf FK, Stukart MJ, Boogert FC, Metselaar K (1978) Limited proteolysis of cloacin DF13 and characterization of the cleavage products. Biochemistry 17:1137–1142

De Vries W, Niekus HGD, Boellaard M, Stouthamer AH (1980) Growth yields and energy generation by *Campylobacter sputorum* subspecies *bubulus* during growth in continuous culture with different hydrogen acceptors. Arch Microbiol 124:221–227

Durkacz BW, Kennedy CK, Sherratt DJ (1974) Plasmid replication and the induced synthesis of colicin E1 and E2 in *Escherichia coli*. J Bacteriol 117:940–946

Ebina Y, Takahara Y, Kishi F, Nakazawa A, Brent R (1983) LexA protein is a repressor of the colicin E1 gene. J Biol Chem 258:13258–13261

Gaastra W, Oudega B, De Graaf FK (1980) The use of mutants in the study of structure-function relationships in cloacin DF13. Biochim Biophys Acta 540:301–312

Hakkaart MJJ, Veltkamp E, Nijkamp HJJ (1981 a) Protein H encoded by plasmid CloDF13 involved in lysis of the bacterial host. I. Localization of the gene and identification and subcellular localization of the gene H product. MGG 183:318–328

Hakkaart MJJ, Veltkamp E, Nijkamp HJJ (1981 b) Protein H encoded by plasmid CloDF13 involved in lysis of the bacterial host. II. Functions and regulation of synthesis of the gene H product. MGG 183:326–332

Hardy KG, Meynell GG, Dowman JE (1973) Two major group of colicin factors, their evolutionary significance. MGG 125:217–230

Herschman HR, Helinski DR (1967) Comparative study of the events associated with colicin induction. J Bacteriol 94:691–699

Horii T, Ogawa T, Nakatani T, Hase T, Matsubara H, Ogawa H (1981) Regulation of SOS functions: purification of *E. coli* LexA protein and determination of its specific site cleaved by the RecA protein. Cell 27:515–522

Hueting S, Tempest DW (1977) Influence of acetate on the growth of *Candida utilis* in continoues culture. Arch Microbiol 115:73–78

Inukai M, Ghrayeb J, Nakamura K, Inouye M (1984) Apolipoprotein, an intermediate in the processing of the major lipoprotein of the *Escherichia coli* outer membrane. J Biol Chem 259:757–762

Isaacson RE, Konisky J (1974) Studies on the regulation of colicin Ib synthesis. II. The synthesis of colIb specific RNA in vivo. MGG 132:223–232

Jakes KS, Model P (1979) Mechanism of export of colicin E1 and colicin E3. J Bacteriol 138:770–778

Jakes KS, Zinder ND (1974) Highly purified colicin E3 contains immunity protein. Proc Natl Acad Sci USA 71:3380–3384

Jakes KS, Zinder ND (1984) Plasmid ColE3 specifies a lysis protein. J Bacteriol 157:582–590

Konisky J (1982) Colicins and other bacteriocins with established modes of action. Annu Rev Microbiol 36:125–144

Kool AJ, Pols C, Nijkamp HJJ (1975) Bacteriocinogenic CloDF13 minicells of *Escherichia coli* synthesize a protein that accounts for immunity to bacteriocin CloDF13: purification and characterization of the immunity protein. Antimicrob Agents Chemother 8:67–75

Krone WJ, Oudega B, Stegehuis F, De Graaf FK (1983) Cloning and expression of the cloacin DF13/aerobactin receptor of *Escherichia coli* (ColV-K30). J Bacteriol 153:716–721

Lau PCK, Rowsome RW, Zuker M, Visentin LP (1984a) Comparitive nucleotide sequences encoding the immunity proteins and the carboxyl-terminal peptides of colicin E2 and E3. Nucleic Acids Res 12:8783–8745

Lau PCK, Rowsome RW, Watson RJ, Visentin LP (1984b) The immunity genes of colicins E2 and E8 are closely related. Biosci Rep 4:565–572

Leduc M, Kasra R, Van Heyenoort J (1982) Induction and control of the autolytic system of *Escherichia coli*. J Bacteriol 152:26–34

Linton JD, Griffiths K, Gregory M (1981) The effect of mixtures of glucose and formate on the yield and respiration of a chemostat culture of *Beneckea natriegens*. Arch Microbiol 129:119–122

Lloubès R, Chartier M, Journet A, Varenne S, Lazdunski C (1984) Nucleotide sequence for the immunity protein to colicin A. Analysis of codon usage of immunity proteins as compared to colicins. Eur J Biochem 144:73–78

Luirink J, Van der Sande C, Tommassen J, Veltkamp E, De Graaf FK, Oudega B (1985) Mode of action of protein H encoded by plasmid CloDF13: effects of culture conditions and of mutations affecting phospholipase A activity on excretion of cloacin DF13 and on growth and lysis of host cells. J Gen Microbiol 132:825–834

Mankovich JA, Lai P-H, Gokul N, Konisky J (1984) Organization of the colicin Ib gene. J Biol Chem 259:8764–8768

Masaki H, Ohta T (1982) A plasmid region encoding the active fragment and the inhibitor protein of colicin E3-CA38. FEBS Lett 149:129–132

Masaki H, Ohta T (1985) Colicin E3 and its immunity genes. J Mol Biol 182:217–227

Masaki H, Toba M, Ohta T (1985) Structure and expression of the ColE2-P9 immunity gene. Nucleic Acids Res 13:1623–1635

Michaelis S, Beckwith J (1982) Mechanisms of incorporation of cell envelope proteins in *Escherichia coli*. Annu Rev Microbiol 36:435–465

Mochitate K, Suzuki K, Imahori K (1981) Amino acid sequence of immunity protein (B subunit) of colicin E3. J Biochem 89:1609–1618

Mock M, Pugsley AP (1982) The BtuB group Col plasmids and homology between the colicins they encode. J Bacteriol 150:1069–1076

Mock M, Schwartz M (1978) Mechanism of colicin E3 production in strains harboring wild-type or mutant plasmids. J Bacteriol 136:700–707

Mock M, Miyada CG, Gunsalus RP (1983) Nucleotide sequence for the catalytic domain of colicin E3 and its immunity protein. Evidence for a third gene overlapping colicin. Nucleic Acids Res 11:3647–3557

Morlon J, Lloubès R, Chartier M, Bonicel J, Lazdunski C (1983a) Nucleotide sequence of promoter, operator and amino-terminal region of *caa*, the structural gene of colicin A. EMBO J 2:787–789

Morlon J, Lloubès R, Varenne S, Chartier M, Lazdunski C (1983b) Complete nucleotide sequence of the structural gene for colicin A, a gene translated at non-uniform rate. J Mol Biol 170:271–285

Nakazawa A, Tamada T (1972) Stimulation of colicin E1 synthesis by cyclic 3′, 5′-adenosine monophosphate in mitomycin C-induced *Escherichia coli.* Biochem Biophys Res Commun 46:1004–1010

Nielsen JK, Lampen JO (1982) Membrane-bound penicillinases in Gram-positive bacteria. J Biol Chem 247:4490–4495

Ohmura K, Nakamura K, Yamazaki H, Shiroza T, Yamada K, Jigami G, Tanaka H, Yoda K, Yamasaki M, Tamura G (1984) Length and structural effects of signal peptide from *Bacillus subtilus* α-amylase on secretion of *Escherichia coli* β-lactamase in *Bacillus subtilus.* Nucleic Acids Res 12:5307–5319

Ohno-Iwashita Y, Imahori K (1980) Assignment of the functional loci in colicin E2 and E3 molecules by the characterization of their proteolytic fragments. Biochemistry 19:652–659

Ohno-Iwashita Y, Imahori K (1982) Assignment of the functional loci in the colicin E1 molecule by characterization of its proteolytic fragments. J Biol Chem 257:6446–6451

Oka A, Nomura N, Morita M, Sugisaki H, Sugimoto K, Takanami M (1979) Nucleotide sequence of small ColE1 derivatives: structure of the regions essential for autonomous replication and colicin E1 immunity. MGG 179:151–159

Oudega B, Stegehuis F, Van Tiel-Menkveld GJ, De Graaf FK (1982) Protein H encoded by plasmid CloDF13 is involved in excretion of cloacin DF13. J Bacteriol 150:1115–1121

Oudega B, Mooi FR, De Graaf FK (1984a) Excretion of proteins by Gram-negative bacteria: export of bacteriocins and fimbrial proteins by *Escherichia coli.* Antonie Van Leeuwenhoek 50:569–584

Oudega B, Stegehuis F, De Graaf FK (1984b) Effect of glucose fermentation on the functioning of protein H in the excretion of cloacin DF13 by *Escherichia coli.* FEMS Microbiol Lett 21:125–131

Oudega B, Ykema A, Stegehuis F, De Graaf FK (1984c) Detection and subcellular localization of mature protein H, involved in excretion of cloacin DF13. FEMS Microbiol Lett 22:101–108

Pugsley AP (1981) Transcription regulation of colicin Ib synthesis. MGG 183:522–527

Pugsley AP (1983) Obligatory coupling of colicin release and lysis in mitomycin-treated Col$^+$ *Escherichia coli.* J Gen Microbiol 129:1921–1928

Pugsley AP (1984a) The ins and outs of colicins. Part I: production and translocation across membranes. Microbiol Sci 1:168–175

Pugsley AP (1984b) Genetic analysis of ColN plasmid determinants for colicin production, release and immunity. J Bacteriol 158:523–529

Pugsley AP, Rosenbusch JP (1981) Release of colicin E2 from *Escherichia coli.* J Bacteriol 147:186–192

Pugsley AP, Schwartz M (1983a) A genetic approach to the study of mitomycin-induced lysis of *Escherichia coli* K-12 strains which produce colicin E2. MGG 190:366–372

Pugsley AP, Schwartz M (1983b) Expression of a gene in a 400-base-pair fragment of colicin plasmid ColE2-P9 is sufficient to cause host cell lysis. J Bacteriol 156:109–114

Pugsley AP, Schwartz M (1984) Colicin E2 release: lysis, leakage or secretion? Possible role of a phospholipase. EMBO J 3:2392–2397

Pugsley AP, Schwartz M (1985) Export and secretion of proteins by bacteria. FEMS Microbiol Rev 32:3–38

Randall LL, Hardy SJS (1984) Export of protein in bacteria. Microbiol Rev 48:290–298

Sabik JF, Suit JL, Luria SE (1983) *Cea-kil* operon of the ColE1 plasmid. J Bacteriol 153:1479–1485

Schaller K, Nomura M (1976) Colicin E2 is a DNA endonuclease. Proc Natl Acad Sci USA 73:3989–3993

Schafferman A, Flashner Y, Cohen S (1979) ColE1 DNA sequences interacting in cis, essential for mitomycin C induced lethality. MGG 176:139–146

Silhavy TJ, Benson SA, Emr SD (1983) Mechanisms of protein localization. Microbiol Rev 47:313–344

Suit JL, Fan M-L, Sabik JS, Labarre R, Luria SE (1983) Alternative forms of lethality in mitomycin C-induced bacteria carrying ColE1 plasmids. Proc Natl Acad Sci USA 80:579–583

Suit JL, Fan M-LJ, Kayalar C, Luria SE (1985) Genetic study of the functional organization of the colicin E1 molecule. J Bacteriol 161:944–948

Suzuki M, Sumi S-L, Hasegawa A, Nishizawa T, Miyoshi K-I, Wakraka S, Miyake T, Misoka F (1982) Production in *Escherichia coli* of biologically active secretin, a gastrointestinal hormone. Proc Natl Acad Sci USA 79:2475–2479

Talmadge K, Gilbert W (1982) Cellular location affects protein stability in *Escherichia coli*. Proc Natl Acad Sci USA 79:1830–1833

Talmadge K, Stahl S, Gilbert W (1980) Eukaryotic signal sequence transports insulin in *Escherichia coli*. Proc Natl Acad Sci USA 77:3369–3373

Terwilliger TC, Eisenberg D (1982) The structure of mellitin. II. Interpretation of the structure (1982). J Biol Chem 257:6016–6022

Tyler J, Sherratt DJ (1975) Synthesis of E colicins in *Escherichia coli*. MGG 140:349–353

Van den Elzen PJM, Gaastra W, Spelt CE, De Graaf FK, Veltkamp E, Nijkamp HJJ (1980) Molecular structure of the immunity gene and immunity protein of the bacteriocinogenic plasmid CloDF13. Nucleic Acids Res 8:4349–4363

Van den Elzen PJM, Maat J, Walters HHB, Veltkamp E, Nijkamp HJJ (1982) The nucleotide sequence of the bacteriocin promoter of plasmids CloDF13 and ColE1: role of LexA repressor and cAMP in the regulation of promoter activity. Nucleic Acids Res 10:1913–1928

Van den Elzen PJM, Walters HBB, Veltkamp E, Nijkamp HJJ (1983) Molecular structure and function of the bacteriocin gene and bacteriocin protein of plasmid CloDF13. Nucleic Acids Res 11:2465–2477

Van Tiel-Menkveld GJ, Rezee A, De Graaf FK (1979) Production and excretion of cloacin DF13 by *Escherichia coli* harboring plasmid CloDF13. J Bacteriol 140:415–423

Van Tiel-Menkveld GJ, Veltkamp E, De Graaf FK (1981) Mitomycin C-induced synthesis of cloacin DF13 and lethality in cloacinogenic *Escherichia coli* cells. J Bacteriol 146:41–48

Van Tiel-Menkveld GJ, Mentjox-Vervuurt JM, Oudega B, De Graaf FK (1982) Siderophore production by *Enterobacter cloacae* and a common receptor protein for the uptake of aerobactin and cloacin DF13. J Bacteriol 150:490–497

Varenne S, Cavard D, Lazdunski C (1981) Biosynthesis and export of colicin A in *Citrobacter freundii* CA31. Eur J Biochem 116:615–620

Varley JM, Boulnois GJ (1984) Analysis of a cloned colicin Ib gene: complete nucleotide sequence and implications for regulation of expression. Nucleic Acids Res 12:6727–6739

Von Heijne G (1981) On the hydrophobic nature of signal sequences. Eur J Biochem 116:419–422

Von Heijne G (1984) How signal sequences maintain cleavage specificity. J Mol Biol 173:243–251

Wagner W, Vogel M, Goebel W (1983) Transport of hemolysin across the outer membrane of *Escherichia coli* requires two functions. J Bacteriol 154:200–210

Walker GC (1984) Mutagenesis and inducible responses to deoxyribonucleic acid damage in *Escherichia coli*. Microbiol Rev 48:60–93

Watson MEE (1984) Compilation of published signal sequences. Nucleic Acids Res 12:5145–5164

Watson RJ, Lau PCK, Vernet T, Visentin LP (1984) Characterization and nucleotide sequence of a colicin-release gene in the *hic* region of plasmid ColE3-CA38. Gene 29:175–184

Yamada M, Nakazawa A (1984) Factors necessary for the export process of colicin E1 across cytoplasmic membrane of *Escherichia coli*. Eur J Biochem 140:249–255

Yamada M, Ebina Y, Miyata T, Nakazawa T, Nakazawa A (1982a) Nucleotide sequence of the structural gene for colicin E1 and predicted structure of the protein. Proc Natl Acad Sci USA 79:2827–2831

Yamada M, Miki T, Nakazawa A (1982b) Translocation of colicin E1 through cytoplasmic membrane of *Escherichia coli*. FEBS Letters 150:465–468

Subject Index

MIX
Papier aus verantwortungsvollen Quellen
Paper from responsible sources
FSC® C105338

If you have any concerns about our products,
you can contact us on
ProductSafety@springernature.com

In case Publisher is established outside the EU,
the EU authorized representative is:
Springer Nature Customer Service Center GmbH
Europaplatz 3, 69115 Heidelberg, Germany

Printed by Libri Plureos GmbH
in Hamburg, Germany